Quantum Energetics and Spirituality

Quantum Energetics and Spirituality

Aligning with Universal Consciousness

Volume 4

KENNETH SCHMITT

Copyright © 2022 by Kenneth Schmitt

All rights reserved. No part of this book may be reproduced or transmitted in any form or by any means, electronic or mechanical, including photocopying, recording, or by any information storage and retrieval system, without permission in writing from the author.

ISBN: 979-8-9851064-3-5

Kenneth Schmitt
Phone: **+1-808-280-4041**
Email: **timeless1@twc.com**
Website: **https://www.ConsciousExpansion.org**

Independently published by the author

Contents

	Foreword	1
1.	The Nature of Our Reality Experience	3
2.	Guidance on the Inner Path	69
3.	Resolving the Limitations of Our Consciousness	125
4.	The Importance of Our Intentions	169
5.	Understanding Our Expansive Self	209
6.	Living in Unlimited Awareness	255

Foreword

The premise of this book is that we are much greater creatures than we have imagined possible. Beyond time and space, we live as our presence of awareness in many energetic dimensions. Although we are all part of the human experience, we each have our unique essence, all participating in the consciousness of the whole and constantly creating the qualities of our own experiences by our mental and emotional expressions.

As we learn to control our thoughts and feelings by choosing the focus of our attention on the qualities flowing to us through the intuition of our heart, we can transform our lives into the experiences we truly want. Nothing exists beyond our own consciousness. Everything we are aware of exists as the choices we make with our attention. We are constantly interacting with electromagnetic waves that we attract and repel by our state of being, manifesting the qualities of our thoughts and feelings, all of which are part of universal consciousness.

Our participation in creation is mandatory, but the quality of our participation is voluntary. We can live in freedom and abundance or in suffering and enslavement. For eons humans have chosen to live in fear. We have been programmed and conditioned by nearly every aspect of society to believe in our mortality and to fear for our existence, constantly struggling for survival and always wary of threats to our well-being. This is all imaginary.

We can know that our awareness is eternal and infinite, and our creative ability is unlimited. We just need to stop sabotaging our creative efforts with our doubts and fixations. Considering how deeply programmed we are, the process of becoming completely positive is a challenge, but it is transformative for every aspect of our lives, and there are many ways of achieving this. We'll analyze the spectrum of our human situation and provide a means for mastering it and realizing our nature as our eternal presence of awareness with infinite creative power in alignment with the life-enhancing conscious life force constantly flowing through the heart of our Being.

This book builds upon the foundation established in the previous three volumes of this series. As we develop our mental and emotional mastery through our realization, we transform our own lives and become powerful transformers of humanity.

Kenneth Schmitt
December 4, 2022

1.

The Nature of Our Reality Experience

Can We Escape from Our Prison of Consciousness?

Moving into the dawn in a new era of history, humanity is being released from the enslavement that many of us were unaware of. Our programming and training have been so thorough, that we have believed that we are free. We have been free only within the confines of the police, the military and the banking and financial institutions. All multi-national corporations, governments and large social organizations have been controlled by dark energy, which is part of the passing era. The truth of our imprisonment is coming into the light, and now the challenge is for us to be ready to walk out the prison door into the light of the new world.

Currently we are seeing the early beams of light seeping through the crack of the opening door. Most are not aware it this yet. But what will happen when the door sweeps open, and we can freely move into a world of light and love? Can we believe it? Can we trust that ascension is real? When all of us are out in the brightness of illumined consciousness, all of the darkest situations and beings that have inhabited them become obvious, and we see that there has been horrendous evil everywhere, feeding from the life force of humanity. These energies are now dissolving in the light and disappearing from our experiences. We can let them go, while maintaining a perspective of compassion for the lost ones and gratitude for our life essence, which shines in the light of Creator consciousness.

The brightness of divine light may be too much for many, and they may choose to stay in the prison of duality, held in by fear and doubt. We need a spirit of adventure to wander into the vibrations of the unknown. What happens when we are loving and compassionate with everyone? How can we even do this, when there are unbelievable threats all around? In the energies of the new era, no threats are real. This was always true, but now it's becoming more obvious.

The power of the dark force has been built on the consciousness of their money, lots of it, stolen through trickery and treachery from all of us, while training us to believe that it is all for our best interests. Now the liquidity of money is disappearing quickly, and the psychopaths are being left to fight over what remains, leaving them without the ability to control us much longer. In response to rising positive energies and higher vibrations, our controllers are becoming desperate and going insane. Many have already left us.

Our world is going through transformative changes that require us to peer deeply into dark energies and release any attachments we have to fear, anger and vindictiveness. We are being challenged to radiate unconditional love, compassion, gratitude and joy in the face of the worst evil. Intuitively, we can understand how the psychopaths took the dark path and became captive within it, unable to escape. Many were oppressed, and they became the oppressors, and the oppressors became the oppressed. We have been through it all. It is time to change our vibrations to life-enhancing energies for all in every moment, carrying us experientially beyond duality.

Understanding Our Lives as Metaphors

When realized from an expanded perspective, our lives are constantly guided toward a more heart-felt expression of ourselves. When we're in tune with our true essence, we flow through life easily, but when we veer too far off course, we struggle against our innate energy. Sometimes we're so unaware, that we need a major event to shake us awake. Everything in our lives has a symbolic meaning for us to realize. When that meaning is of great importance, and we're not getting it, we face great tragedy, making it impossible to continue as we have been.

Every message to us is all around us, waiting for us to realize its meaning. If we understand our lives as symbolic, with much deeper meaning than we have realized, we can search for that meaning everywhere. When we search for it, it comes to us through the energy of the heart of our Being, our own inner knowing, and it encompasses our entire awareness. It is the dawning of our realization of who we really are and what we're doing here.

We are not mere humans as we have thought of our species. Humanity is our current personal expression, but we are so much more. Each of us has to know this for ourselves. It happens when we're open to it and receptive without limits. Our self-imposed limiting beliefs about ourselves have kept us from realizing our expanded reality. By practicing being focused with our attention, we can control our thoughts and emotions so that we can be present in awareness in every moment. In this state of being, we can be open and receptive to our inner realization.

No longer do we need to entertain anxiety in our ego-consciousness, because our natural flow is easy. Our gratitude keeps us aligned with our intuition. Gratitude has an energetic level that attracts resonating energetic patterns that we realize as our wonderful experiences. Once we enter this path, and until we are completely transformed, we will experience the results of our miss-directions. If we're strongly attached to a limiting belief, we may encounter a great challenge that requires a new perspective. If we can accept this, the way is made clear for us to proceed through our transformation leading to our realization of eternal, infinite presence of awareness in unconditional love and joy beyond and including our human consciousness, and with infinite creative ability.

We are a species with this potential level of consciousness. We can learn to open ourselves to its reality by developing acute sensitivity to our inner realization. In this state of Being, we allow ourselves to be guided by everything around us and within. We are not required to participate in negative experiences at their energetic level. When we can maintain alignment with gratitude, we are beyond the influence of negativity, and our experiences align with our energetic state of Being.

Preparing Ourselves for Our New Reality

In recent years our quality of life has taken a significant diminishment at the hands of the global elite. We have barely escaped terminal enslavement, and now the entire Matrix is coming apart and dissolving. What will replace it, and how will we survive without our psychopathic rulers? We do not need to know the details, or even the forms that will arise. We only need to be in gratitude and to know that we are always abundantly provided for. This would already have been our experience, except that we gave our divine inheritance away to the thieves and parasites by aligning ourselves with their energy, either by submission or rebellion.

Chapter 1. The Nature of Our Reality Experience

The new world is an experience of love, joy, abundance and freedom to do and be whatever and whoever we want, apart from interfering with another being, including the Earth. This may be beyond belief for many, considering our current predicament, which appears to be getting worse. It is only the dying throes of the parasites, as they thrash and scream insanely for their final apocalyptic moments, desperately trying to steal our life force through fear and intimidation. We are not required to give it to them. We have free will to choose how we think and feel about everything. No one can make us react to anything in fear or dread.

We are the creators of our lives, and we always have been. By our naivete and acceptance of negative energy, usually disguised as positive by our rulers, we have allowed them to suck the life force out of us through taxes, currency manipulation, military conscription and warfare, interest charges, intentional plagues, poisoning of our food, water and air and enforced injection of toxic and lethal substances. But if we knew how to control our own energetic alignment, we never had to accept any of this.

The appearance of physical solidity has beguiled us into thinking that the empirical world is our ultimate reality, when we now know through quantum experiments and mathematics that everything is energy, and materiality is only apparent. We are energetic beings. In fact, we are energetic modulators, able to control the quality of our experiences through the polarity and vibratory level of our imaginary visions and feelings, as we realize the light within all beings and situations.

By aligning ourselves with the vibrations of vitality and transcendence, we attract only experiences that are in resonance with our state of being, regardless of the negative energies swirling around us. Positive energies are always present for us to experience, which becomes possible when we recognize them and align with them. As Jesus said we could, we now know how to do everything he could do and more.

Being Human in a Time of Expanding Consciousness

In this most crucial time in human history, we are facing a complete transformation in our way of living and knowing our true identity. All of the structures and operations that we have depended upon and interacted with are breaking down. Our society is in chaos, our monetary system is beginning to fail, our governments are becoming tyrannical and destructive, and anger and violence are prevalent.

Our conscious awareness has been limited by the belief that we exist only in an empirical world of time and space. For this experience, we created and developed our ego-consciousness, defined by our limiting beliefs about ourselves. Many have believed that we are on our own here, not knowing how we got here, or whether there is a purpose for us and a meaning for our lives.

We've always had spiritual masters who were not subject to the same laws of the material world that everyone else is. The masters could expand their consciousness beyond the body and open their awareness to distant realms. They could change things in form and create whatever they wanted. They taught that all of us have these abilities, when we are open and receptive to our true Self. They taught that we are all the same Being with great creative ability.

With the rise of quantum physics, and given the use of appropriate technology, anyone can prove that consciousness is universal and is the cause and basis of everything. We can logically deduce that universal consciousness is the expression of a conscious Being, who directs the existence and essence of everything. Everything exists within the consciousness of this Being, including us. Everything participates in consciousness, including things that we have regarded as inert.

What does it mean that we participate in universal consciousness? In our potential realization, we are unlimited in awareness of the nature and quality of the Being who creates us, and we share in unlimited creative ability. We can potentially feel and know the nature of the Being whose consciousness we arise out of, and we are free to use our consciousness however we want.

In the past, our cosmic environment was polarized for duality of light and dark, life-enhancement and life-diminishment. As we enter a new round of precessional planetary motion, the energetic resonance of the Earth has been changing toward only positive, life-enhancing energies. In expressing the unconditionally-loving feelings and thoughts of our Creator, we and our planet are evolving into greater awareness and conscious expansion of creative ability, in unity with all conscious life. We are designed to be creative in every way.

Elevating our Reality

We inhabit a realm that has an awareness threshold that keeps humans from recognizing other realities that interpenetrate ours on an energetic level. This threshold is within our consciousness and consists of our limiting beliefs about ourselves and about the nature of our reality. We have been choosing to focus our attention on one of the many channels in our potential awareness. If we desire, we can change channels and be in the awareness of another dimension of the expression of consciousness. If we shift from duality, where there is a mix of positive and negative energies, into only positive polarity, our entire lives change.

This is such a major change, that it requires a lot of practice. Because we must live in the moment, we do not look for assurances outside of ourselves, and we need to be able to trust our intuition. By becoming sensitive to our inner knowing, we can access more assistance than we need. We have guides and angels, ascended masters and higher beings who offer assistance through the energy of our heart.

If we do not align ourselves with any negative energy, which has an element of fear, we do not feed it our life force through our attention, and it disappears from our experience. It may still be happening around us, but it's not in our personal lives. While we stay focused on positive, life enhancing experiences, we elevate our energy signature, and our radiance affects others around us. We attract those who resonate with our vibratory alignment.

Once we learn to depend entirely on our intuition, we can resolve any remaining limitations hidden in our subconscious self. They dissolve, because they cannot exist without fear. Fear is absent in a positive perspective. There is only the level of joy, compassion, love, abundance, beauty, freedom and sovereignty. These are the qualities of a higher dimension that we can live in here and now, and they are our natural state of Being, when we are heart-centered.

Necessarily Transcending Ego-Consciousness

For eons our ego consciousness has guided us through the world of duality, believing that we are separate beings from one another, subject to the actions and controls of others outside of our own consciousness. Even now we wonder about our future. Will the monetary system crash and our money become worthless? Will there be expanded warfare? Will we be poisoned by the pharmaceutical, pesticide, herbicide and GMO companies? Will we become transhumanized by artificial intelligence and lose our free will? How can we withstand all of these outside threats to our well-being?

The big question is, how do we know any of this is real? It is real, if we believe that it is. What if we choose not to believe it? Everything about the world that we experience as humans is a complex, interacting matrix of energetic patterns held in the consciousness of humanity as reality. Only can a powerful mass of awakened individuals change the conscious fixation of humanity, and this is transpiring, but we as individuals can change our own perspective and transcend the experiences of humanity.

The quantum sciences of the past century have given us some wonderful insights about the nature of our reality. Things are not as they appear to most of us. If we reduce everything to a subatomic level, it becomes clear that every electromagnetic wave and pattern of waves exhibit conscious awareness. Energetic patterns that we recognize become empirical entities when we recognize them. They are also multidimensional and can exist in more than one place at the same time. From this we can deduce that they are expressions of consciousness, and that every aspect of ourselves is consciously aware.

What does this have to do with our predicament as humans? We have the ability to control our consciousness. We can accept into our experience only the interactions that we desire. Just as our constituent parts are held in our consciousness, we have the ability to change the quality of their vibrations by our state of being. Everything in our bodies, our relationships and our experiences is a result of our beliefs and perspectives. The quality of our conscious awareness controls everything in our lives.

Our ego consciousness cannot conceive of how this could be, but intuitively we can know and align ourselves with the vibrations that are most life-enhancing and empowering. Living in this knowing and feeling the energies that we are most attracted to results in living the expression and experience of these energies. Our life experiences come from within our own consciousness.

Our Contributions to Divine Consciousness

We are experience creators within the consciousness of the Creator. Every experience in our awareness is in the universal consciousness of the Creator. Our human consciousness is confined to a self-limited spectrum of dualistic energetics. This is a clever game that we participate in, and we constantly contribute our experiences to universal consciousness. Aside from this game, these experiences would not be possible for us or any higher being to know. We have experienced our limited identity so intensely, that we have felt and believed ourselves to be separate from the consciousness of the Creator, even though this is impossible, or we would not exist. We know this innately, and quantum physicists have also concluded that universal consciousness is the source of everything within and beyond space and time.

We exist within and beyond space and time in our conscious awareness. Our true awareness is infinite, as is our creative power. We are the ones who think, feel and sense experiences for universal consciousness. We have complete freedom to create energetic patterns with our state of being, our thoughts and feelings. We are players on the Earth stage, and we create the quality of our experiences just by how we believe ourselves to be. We also have the use of our intentional focus of attention and our ability to sense resonances and align with compatible energies. We can choose what we want to pay attention to, and we can use our attention intentionally in creative ways.

By being in a positive state, commanding positive emotions, envisioning and thinking of life-enhancing scenarios, we create that energetic level for our experiences. The intentions that we create are universal. This is the reason that we naturally want life-enhancing experiences for everyone. Our intuition guides us to interact with others from this perspective. We can be entirely Self-motivated, Self-sustaining, and unconditionally loving of all life and everything in eternal gratitude. We are fractals of the Creator, existing eternally in universal consciousness. This awareness is available to us now.

Our Self-Expressions and Potential Awareness

Everything around us is a reflection of our own energetic creations. If we have been living under the guidance of our ego-consciousness, it is all a distraction from awareness of our true Being. We have kept ourselves interminably busy in patterns of survival, so that we have been unable to confront the truth about ourselves. We have not known the essence of who we are, or what our true abilities are. Our ego needs protection, care and adulation, because it is an artificial creation without its own life force. It exists by our recognition of it, and our belief in it as real. We have been trained to believe that we are our ego, empowered by our mysterious subconscious, and in extreme situations, guided by our superconscious or higher Self.

To regain our connection to our true Self, we can evaluate the energy that we encounter and that flows through us. We can begin to determine its polarity. Is it negative, expressing some form of fear, or positive, expressing joy and life-enhancement? By moving away from fear and toward appreciation and gratitude in our own perspective in every situation, we can begin to transform our life experiences.

Nothing comes into our experience, unless we recognize it and believe in its reality, giving it permission to enter our lives. The ruling elite have become masters at training us to comply with their program and accept our enslavement. We are now in the final act of this play, and our choice is either ultimate submission to transhumanism and complete enslavement or defiance and freedom to live in unconditional love and alignment with the consciousness of our Creator. The energy patterns for both of these scenarios have been intersecting for some time, but then began diverging. We are being compelled to choose the path of our destiny. The experience of duality or eternity.

The Necessary Ingredients for Creative Living

Most of humanity believes that what we experience through our senses is real, and that our real creative power is what we do physically in the world. This ignores what we know from quantum physics and spiritual teachings. The empirical world is our conscious interpretation of energy patterns that we recognize and make real for our selves as a result of our telepathic alignment with human consciousness.

We can project ourselves into any social situations that we desire. It requires our alignment with the same quality of vibrations that we desire. We have the power of our imagination and intention to direct our vibration to whatever level we pay attention to. Our subconscious innate being has a strong influence on our experiences, and we must communicate with this part of ourselves with love and gratitude. This is the level of resonance where alignment with our innate being is possible without limitations.

Because everything is electromagnetic energy patterns held in empirical manifestation for us by our consciousness, we can change those patterns by our intentional alignment with our natural energetics, which we know by how we feel most fulfilled, most compassionate and joyful. Once we know with confidence in out intuition that in our true Being we are our eternal present awareness, with unlimited creative power and ability, we have transcended belief of the ego and are free to expand our awareness into universal consciousness.

In our current experience as humans on this planet, we must participate in a matrix of duality, but we are not limited to this level of consciousness. We have the ability to transcend duality by alignment with positive, high vibrations. We can shift our consciousness out of the realm of doubt and fear and into the realm of love and confidence. We don't have to change anything in our lives, unless we want to, and we are constantly guided intuitively when we are aligned with the energy of our heart.

It is strong and unconditionally loving, and it can guide us to everything our heart desires.

Using Quantum Energetics for Transformation

Because everything in our experience is energy in its myriad patterns and frequencies, we can train ourselves to be sensitive to the polarity and vibratory level in our awareness in every moment. Whenever we feel the negative polarity of fear, we feel threatened. This is part of our experience in the realm of duality. How we react to it is important. Do we believe that we want to resist it, accept it or transform it? Resisting it gives it reality for us, but we are engaging it on its own level and feeding it our life force through our alignment with its polarity. If we accept it, we acknowledge its reality for humanity, but we do not need to engage with it or give it our life force. Then we can transform it in our personal experience by being in a positive state of compassion and joy.

When we are thoroughly focused on being filled with light, love and joy, we are not subject to negative energy patterns. They may be present all around us, but they cannot enter our personal conscious presence without our permission. Everything in existence requires conscious creative support, which flows from the unconditionally loving essence of the Creator, and, in our experience, through us. We are fractals of the universal consciousness of the Creator, and we have the free will and the ability to modulate positive and negative energetic patterns into physical experiences through our state of being in our thoughts and emotions, which we can learn to control.

Although we are part of humanity, and humanity holds the world of duality in conscious awareness, we have the free will ability to have our own experiences. When we identify only with positive vibrations, and we radiate kindness and life-enhancing thoughts and feelings to everyone and everything, we gain the ability to transform all of the energetic patterns that we encounter into alignment with the universal consciousness of the Creator. We cannot change the polarity of other beings, unless they want to align with us, but we can offer them support to be in joy with us.

When we are in a positive vibratory state of compassionate understanding, we can claim our personal sovereignty and exemption from negative influences. We can be aware of ourselves as being pure, unlimited, present awareness, expressing the life-stream of Creator consciousness in beauty and freedom through our physical embodiment. In this way we can transform and elevate all life on our planet.

The Richness of the Void of Ego-Consciousness

From a place of nowhere, we are present awareness. Our awareness requires only the flow of conscious life force from the Creator. This life force provides an open connection through our intuition, our inner knowing, into the consciousness of the Creator. We are endowed with personal Self-Realization that is unlimited in universal consciousness. To experience it as our reality, we must recognize its energetic resonance and align with it.

The transformation of the Earth provides us with some clues. Geophysicists are observing the weakening and changing of Earth's polarity into a status of unitary polarity within the center of the planet. Gaia is shifting to only positive on every level. Negative energy will soon be unable to exist here. All beings will be positive beyond polarity. The negative has been getting weaker and disappearing for a decade now. It still appears to be powerful, but it is only appearance, because of the media.

If we do not engage with negative vibrations or align with them in any way, they cannot come into our experience. We can be subject to psychic and spiritual parasites only with our consent, in which case we must align with their energetic resonance. Only if we give them reality by believing that they are real for us personally, will we have experiences on their vibratory level. By vibrating completely positively at a level of gratitude, compassion and love, we remain beyond their influence and encounter.

It is our awareness in every moment that establishes our limits or our infinity of Being. We are either open or partially-open to our true Being. We cannot be completely closed, or we would not exist. Everyone has inner light. The more light we can be open to, the more expansive we can be, and the higher our vibrations. All of our cells receive conscious life force and emit photons. We are radiant light Beings, even in physical form.

Our bodies are symbols of what we believe about ourselves. They change as our beliefs change, and as our relationship with our innate being changes. Our innate makes no judgment, it just expresses our conscious energetics as physical features and conditions in our bodies. Being observant about this is how we can understand our limiting beliefs, so that we can resolve them in order to become unlimited in our awareness.

Choosing the Quality of Our State of Being

As humans, we do not really know how to love. We have been programmed and taught to believe that we are not worthy, that we are inadequate. To understand unconditional love, we must align our awareness beyond the world of duality to our true intuitive knowing. Love, joy and gratitude come to us through our intuition and all vibrate together in resonance. They are the emotions of our true Self, and it is through calling them into our lives that we can come to know our expanded awareness.

If we desire to, we can align with our intuitive guidance, which comes to us on every level of our being in ways that we can be aware of. It comes through our heart constantly in our conscious life force, arising from the consciousness of the Creator. Our essence is cosmic consciousness and beyond, but we have been blocking ourselves from realizing this. We've had to do this in order to have a realistic experience of living in duality. There's no requirement that we restrict ourselves to this realm. We are much greater and can choose to realize our true Being.

We can realize that all of our experiences are projections and reflections of our own consciousness. We live in a kind of virtual reality held in awareness by the consciousness of humanity. As humanity changes its awareness, the energetics of our human experiences change. When we learn that our mental and emotional states are conscious choices for us, we can begin to realize our true Self. Although we are each part of humanity, we each have absolute control of our energetic presence.

The controllers of society can direct our experiences only with our permission. We are not required to align with their energy. We can move beyond their negative polarity of fear and servitude into an experiential realm of joy, love and gratitude with confidence in our unlimited abilities. When we can trust ourselves to be receptive to the life-enhancing guidance of our intuition, we are infinitely powerful creators, with nothing more than our conscious presence.

No one else can experience what each of us experiences. Our entire human experience consists of recognition of patterns of electromagnetic waves that stimulate our senses, and that our consciousness interprets as the world of our human lives. We are free to focus on any polarity and frequency that we choose, consciously or subconsciously. The vibratory quality of our attention is what we energetically align with. It is what we project into the quantum field and what is reflected back to us as experiences in the world we recognize and believe as real.

Chapter 1. The Nature of Our Reality Experience

Living Beyond Polarity

The leap in consciousness required in being only positive, is beyond the limits of ego-consciousness. Within the larger spectrum of humanity, it is the unknown, and therefore feared and guarded against. We have powerful limiting beliefs, but we can align with and rely on our intuition. Intuitive knowing is what we need most to guide us into higher vibrations. Once we know ourselves in our eternal Being, our limitations drop away from our reality. We can recognize them for what they are, but we can align with positive vibrations and intuitive knowing of everything relevant to us. We can be the observer of ourselves from beyond ego-consciousness. We can know the vibrational script of our human lives and know the feelings that we experience, while we can align with our greater Self through our intuition.

Our lives can become stress-free and filled with love just by our vibrational alignment in our perspectives and feelings. When we are filled with gratitude and joy, we create experiences that stimulate those vibrations within us. We do this by being the one that we want to be, by training our thoughts and emotions to be positive as much as possible. We can be aware of the entire human experience from a positive perspective of constant guidance toward inner knowing.

When we're aware that we've been living within a compartment of consciousness that is our own constant creation and limitation, we can understand from a perspective of compassion and love what our human life is about. The difficulties we experience come about in order to guide us to want to become more positive and loving. This is the quality of the energy that enhances life. We are being drawn to being positive in everything. It's all clear on the level of energetics. In order for creative life force to express itself, there has to be only positive polarity in everything, because the negative is destructive. Positive energy is the primary influence in the conscious life force of everything, and when we are in alignment with this energy, we thrive with vitality.

The universal consciousness of the Creator is beyond polarity, because polarity is a concept limited by time and space. Universal consciousness is entirely creative of life. In the perspective of duality, universal consciousness is only positive. When we can imagine living in a realm of love and joy, and we can feel ourselves being in that vibratory resonance, we can begin to realize its reality. It is a dimensional leap concurrent with our normal experience. The difference is that one path that we recognize as real is guided by the ego, and the other is guided by our intuition. These paths are moving farther apart energetically as the polarity of the Earth shifts to mono-polarity that is positive in the center.

All life on Earth is moving through a dimensional shift in conscious expansion, and the most powerful vibrations enveloping us are stimulating a desire to be positive in every way. Being present in awareness in nature can be a wonderful way to connect with the rising vibratory energy of Gaia. We can ask to become aware of our inner light and intuitive knowing. We can practice being sensitive to these energies. They come from the heart of our Being and are what we know innately, beyond limitations.

Enhancing Our Personal Encounters

We may feel that we want only loving and joyous encounters with others. We can realize this quality of experience by aligning ourselves with positive, high-vibratory energy patterns in our emotions and imagination, even when facing challenges. If we can intentionally respond to every encounter, be it positive or negative, in a positive, life-enhancing way, we are opening our awareness to the reality of a higher spectrum of living. Our personal encounters become filled with gratitude and joy. Here our friends are true, as are we. We are Self-sustaining and supportive of one another and can be aware of the exchange of energies in every encounter.

What is an encounter energetically? It is two persons recognizing each other's energy signatures. The wave patterns of energy in our personal expressions manifest as our bodies and personalities, but we are encountering energetics that stimulate feelings in us, as well as mental interactions. In universal consciousness, all interactions are based in love. Fear does not exist. Currently it can exist for us, because we have the free will to create it by compartmentalizing our conscious awareness and polarizing ourselves with negative energies as well as positive.

Within the compartment of human consciousness, we can experience things as real, that we could not believe could be real in our eternal, expanded awareness. We have actually expanded universal consciousness to include the qualities of negative energies. We have deepened the understanding and compassion of all beings on an energetic level. Because we express ourselves as energy beings, we can align ourselves with energy of any polarity or vibratory frequency that we choose, resulting in experiences for us of those qualities.

We can imagine scenarios in which we are living the way we really want to live, and we can feel ourselves in that virtual reality. When we can maintain that level of thinking and feeling as much as possible, we train ourselves to realize its reality. Anything we can imagine already exists in the unified quantum field. We only have to align ourselves with its energetics to experience it.

When we are primarily interested in meaningful and loving encounters, we are aligning ourselves with the life-enhancing intent of Creator consciousness and expanding our awareness beyond human limitations to intuitive knowing.

In our human separation from each other, we have been able to experience what it is like to encounter another being on a physical level. Energetically, these are valuable experiences that would not have been possible in our expanded consciousness. Once we understand our true, unlimited Self, we can be the energetic masters of our lives and truly enjoy being human.

Managing Injuries and Pain

As long as we are human, we will continue to witness negative, constricting and self-destructive energies. When we are aligned with our intuitive guidance, we can receive them with compassion and understanding without aligning with them. If we intend to align only with positive vibrations, for a while we may still encounter negative experiences, including personal injuries, physical, psychological and emotional.

When we are on the path to personal expansion, our injuries are all designed to break the hypnotic trance that we have been in and to draw our attention to what is important in our lives. Their symbolism can give us insight into their message. If we can align with our intuition at these times, we will know how the encounter is important for us. These are all things that we must learn for ourselves, and we can have all the help we need, if we are open and receptive to it. This is true in every aspect of life.

Injuries and pain are a result of negative energy that we have aligned with at some time, or even inherited. It is all misdirected energy containing a message for us to come into greater alignment with our higher guidance. It can be very specific or more generalized, and it's the kind of experience that we can understand, if we want to. Our pain may last as long as we need to reorient ourselves to a more meaningful life.

We can receive our injuries and pain with gratitude for awakening us to something important that we have been resisting. Once we recognize the message, we can forgive ourselves for getting stuck in negativity. When we can change our perspective to positive, we can proceed to heal and begin to live at a higher vibratory level.

In order to feel filled with vitality, we can be positive and helpful, allowing our life force to continue to flow and open our receptivity to greater life experiences. When we remain open and receptive to love and gratitude, we continue to expand our awareness and our vitality, even in the face of deeply-negative encounters.

The Origin and Destiny of Our Ego-Self

Our ego consciousness is an extension of our true Being, but conceived in limitation. We wanted to experience living in duality, and we chose to limit our consciousness in convincing ways, so that we could believe our human experiences are real. When we choose to resolve our limitations, the ego returns its life force to us. Henceforth we can be guided by our intuitive knowing. Resolving our limitations is possible with higher guidance, which the ego has no awareness of. Although we have deeply believed that we are our ego, we are much greater, and we can open ourselves to this realization.

Only our doubt and fear of failure keep our creativity limited. We can create instantly when we know that we are the infinitely powerful Creator, far beyond the limitations of humanity. We can align ourselves with the most positive, highest vibrating feelings that are possible for us. This level of energetics is where we can most easily be aware of our intuitive guidance and inner sound current, which disappear from our awareness when we are not paying attention to them.

Developing precise sensitivity to our intuition makes it possible for us to transcend ego-limitations and expand our awareness beyond the duality of the compartment of human consciousness. We can enter the world of only positive energetics, which we feel as love, gratitude, compassion, freedom, and joy.

We can feel completely supported by our unlimited creativity. Without doubt in our abilities, we can live from the love and vitality of our heart in our encounters with others and with nature. All of our intentional creations manifest in our experience with the quality of our state of being when we conceive of them. When we imagine living in gratitude and joy, we are aligning with the energy of life-enhancement arising from the consciousness of the Creator. We can open our awareness to universal consciousness.

In our true Self, we are unlimited in awareness and creative ability. We define ourselves by the energetic levels that we choose to pay attention to. It is possible for us to be aware of our greater Self. This completely transforms our human experience and enables us to be fully-conscious creators of the quality of our lives in every aspect. The ego-mind disappears from our lack of attention to it, and we no longer need to think about things. We can be our present awareness, knowing and understanding the present moment always, while going about our lives.

How We Can Love Our Bodies

Our bodies have abilities that connect us to the natural world in subtle ways. Because we are all the same consciousness, we can enter the awareness of other beings. We can feel their radiant presence. We can align with their vibrations and feel their essence, and we can do this with the Spirit of the Earth.

We may have beliefs that block this awareness by convincing us that trees and other plants are just things without consciousness, and that animals are separate from us and have no conscious identity. Likewise for our body, but we can decide to open our awareness to it as a vibrant vehicle for us to be able to experience everything that only physical human life can provide.

The condition of our body is a result of the continuous presence of conscious Creator life force enlivening our body, with the limitations that we have placed upon ourselves and accepted through inheritance. These personal beliefs are needed for living in duality. They exist, because we have created them to provide a convincing experience here. They are the cause of all of our physical defects, which are signs for us to resolve our attachments to negative energy, all of which are life-diminishing.

When we are holding onto all of the energies embodied with our ego-self, we cannot allow our body to regenerate to its divine essence. To realign our physical energetics with our divine blueprint requires recognition and resolution of our limiting beliefs. Any attachments to negative energy keep us from being able to be only positive.

We often align with negative energy by engaging with it on the level of resistance, which is life-diminishing and draining of our life force through anger and fear. The solution is to transcend the negative energy by intentionally and willfully aligning with positive vibrations. This takes us into a higher dimension of living. If we can be positive always, we have no attraction for negativity, and we do not experience it, regardless of what is happening around us.

As we become more loving, compassionate and open to higher positive vibrations of all kinds, we can also grow in accepting and loving our physical presence, and we can be grateful for the experiences that our body provides for us. We can know how best to help our body to function and serve us. We can give it the stretches and exercise that it needs, nutritious food and healthful care. Our body consciousness controls every cell and molecule in our body, and it operates in alignment with our state of being, the polarity and vibratory level of our predominant thoughts and feelings. By being in love, gratitude and joy, our body naturally regenerates and becomes more vibrant, once we have resolved our deep-set beliefs that doubt that this is possible.

Once we become mostly positive, and we have begun to realize our intuitive knowing, we can be mostly loving and supportive of the positive expression of everything, including our bodies. Our body consciousness is part of our Being. It knows when we are positive and aligns with this energy, providing life-enhancing vibrations for our body.

In our true Selves, we could never take the limitations of humanity seriously, because we realize our present, unlimited awareness, living in the eternal consciousness of the Creator, and able to create everything we need and want. Awareness of our expanded Self can allow us to play our roles in the Matrix of humanity with the mastery that our intuitive guidance provides in alignment with unconditional love for our entire Being and all that exists.

Knowing What's Real

With all the chaos in the world, we may be easily distracted in believing that we are victims of our circumstances, especially if we are in a combat zone or are starving and seem to have no means to excel in our lives. It can be helpful to realize what actually is the cause of the qualities of our experiences.

Most schools of philosophy posit that the empirical world is our reality, and that we can know only what we can define logically and experience physically. There is little understanding of the nature of consciousness, but what happens when our bodies die? Our conscious awareness does not disappear. Instead, it expands greatly, and we become unlimited in our presence of being.

If the empirical world can disappear from our conscious experience, how can it be our reality? The only permanent experience that we have is our conscious present awareness. Even when humans are in a coma, our awareness does not disappear, it just resides outside the body. Without body consciousness, we still feel our encounters with each other's radiant presence, and our personal consciousness expresses itself in our energetic signature, but in a different dimension of being.

In a non-physical dimension of living, we still experience the qualities of our state of being and our encounters with others as expressions of our polarity and level of conscious vibrations. We can sense the energy in the heart of our Being, just as we can when we are embodied, but without the density of physical embodiment.

Our physical embodiment is valuable for experiencing the results of our energetic creations. Because we are masters of modulating patterns of electromagnetic waves that our consciousness interprets as empirical experiences, we have the benefit of experiencing in physical form the results of our thought-patterns and emotional states. As we gain awareness of our preferred experiences, we can learn to direct our thoughts and emotions in alignment with the unconditional love and joy of the creative consciousness that is the Source and essence of every conscious being in every dimension. When we align ourselves with this level of energy, our experiences come into resonance with us, and our reality transcends dimensions.

Deepening Our Understanding of Life

In the eternal continuum of now in the quantum field, every energetic pattern that has ever existed and that ever will exist is present for us to recognize and know through our intuition. Because there is only one unified consciousness, it is all contained within our own potential awareness. In order to participate in the limitations of human experience, we have veiled ourselves off from our unlimited awareness, and we have not realized how we control the quality of our lives.

Our deeper understanding depends upon developing sensitivity to our intuitive knowing. We can do this by asking our spiritual guides and angels to prompt us in this direction. We can ask to receive guidance in the best ways that we can recognize and understand. If we are out of touch with our emotions, we may receive visions and mental constructs and symbols, but ultimately it is our emotional sensitivity that is most important, and that we can feel and trust for guidance and knowing, once we have developed emotional clarity.

The mind can be easily distracted and confused, but our emotions are solid. When we genuinely feel something, there is no question in our awareness, and our mind can follow. Our emotions are sensitive to the radiance of all beings and their quality in our presence. When we want to know or experience something, even something that is hidden, we can imagine the feeling of being in the presence of what we want and how we think about it. We can imagine that we're experiencing it now in gratitude and joy.

Whatever it is, it is already available to us in the quantum field. It does not matter if it has never been experienced in human life. If we can imagine it, it already exists and can be called into our experience. This is possible because in our essence we are multidimensional, eternal awareness. The only thing that can keep us from realizing this is our limiting beliefs about ourselves.

As we are able to resolve and transcend our limitations, a much greater life opens for us. There is nothing outside of our own consciousness that can limit us in any way. The way of mastery is aligning ourselves with the creative, life-enhancing energy that flows through our heart as our intuitive knowing and the source of our consciousness and physical being.

Facing Our Destiny as the New Humans

We are evolving into a complete change in our way of living, and we are being directed to pay attention to our inner knowing. Everything outside of ourselves is becoming unstable and challenging, and as we move forward, this will intensify. The old controllers of our world are disappearing. All of our governing institutions are dying, and our societies are experiencing rising

Chapter 1. The Nature of Our Reality Experience

chaos, shortages of everything, and political insanity. This is getting more intense.

We are experiencing the turning of the ages, and the old energetic patterns are dissolving as the new expressions of a higher vibratory life come into our human experience. The more we hang onto the old ways of depending on sustenance and support from outside of our own being, the more uncomfortable we will become.

This is a time to learn how to understand life beyond our ego-consciousness. We are being challenged to seek higher guidance within our own Being. The transition from the old ways to the new may be rocky for many, and the more we resist, the more we will suffer. There is no place to hide. The galactic energies are penetrating everywhere. We will have to rebuild human life on this planet from within our intuitive knowing.

Once we realize what is happening, we can begin to align ourselves with higher vibrations of compassion and love. These are the energies of the new world. In this perspective, we can open ourselves to what we truly know about our life. There is unconditional love and abundance for all, if we are open to it and confident in receiving it. Without doubt or fear, we can change our perspective to being only positive.

Even though we may not have realized it, we are the masters of creating our life experiences in every moment. By our state of being in our thoughts and feelings, we attract the compatible energetics that become our experiences. If we have not learned this, we are entering the process of being required to. We can learn to clear ourselves of attachment to any eventuality or limitation and open ourselves to the truth of our Being, which we can know in the intuitive energy of our heart.

Moving Toward Personal Transformation

The elite who have ruled this world have developed a masterfully intensive program to blind us to our true identity and use our own consciousness to make us subservient. Nothing outside of our own consciousness can affect us in any way, unless we create it. Everything that binds us is real for us only because we recognize it as such and believe in its reality. We are its creators, intentionally or by submission. The powers that ostensibly control our lives receive all of their force from us. Because we believe that we have limited ability, we feel that we must align with their negative polarity through support or resistance. Through our fear and submission, or anger and resistance, we give them the life force of our divine consciousness. This weakens us and deprives us of our greater creative ability.

Once we realize our situation, we can begin the process of remembering the truth of our Being. We can develop sensitivity to our inner knowing and become intent on thinking, feeling and acting on what feels most uplifting and enhancing for all life. This knowing comes to us in every moment. As we continue to develop our inner sensitivity, we can learn to become positive in every way, raising our vibrations in alignment with the emotions of our true Self. In joy and gratitude we can begin to realize personal responsibility for our lives, resulting in our sovereignty and freedom.

Assisted by the rising positive energies of the Earth and the glowing, high-energy photon cloud of our galactic environment, we are being prompted to free ourselves from our historical bonds of enslavement and open our awareness to the potential of universal consciousness and mastery of our lives. Many techniques are available to help us center our awareness and transcend our ego-mind in deep meditation, where we can realize our true Self.

Spiritual progress requires intentional practice and motivation, and our current conditions are more supportive of this than ever before. As the Spirit of the Earth centers herself in positive polarity, we are increasingly supported in feeling the elevating energies of gratitude, compassion and joy, and it becomes more difficult and uncomfortable to be negative and controlling. Our planetary environment is becoming brighter and more beautiful, and we have the opportunity of aligning ourselves with these energies. We are here for a most wonderful transformation of life as we know it.

Developing a New Perspective

We may realize the fullness of who we are as pure awareness beyond time and space, knowing our own identity and realizing our own energy field, within the all-encompassing consciousness of the Being who creates everything.

For our ego-consciousness, this perspective is impossible to believe, because the ego has no access to higher guidance. Our ego needs limits and separation from the Source of our Being, always seeking security and being in some level of fear. We have been afraid to give to others beyond our perceived means. We've been afraid we might get sick or hurt, afraid or angry that we may be threatened or further enslaved. As long as we have fear, we are in our ego-consciousness and are terminally limited in our expectations.

Being in fear is a choice. It is not required. It is only a reaction, and when it is maintained, it becomes a state of being, even on a low level. Fear is the opposite polarity from love, and it makes love impossible, because we cannot be negatively polarized and be able to act in a positive way. We can pretend and even convince ourselves that we're being positive, but if we have any fear, love is impossible.

It is a great challenge to choose to be only positive, and to transcend ego-consciousness. It requires a powerful intention and a willingness to enter the unknown realm of love without fear or doubt. It is beneficial to develop sensitivity to our intuitive guidance, which is our inner knowing of what we are being drawn to pay attention to, and how to feel about it and imagine it.

We can confidently assure our ego-identity that everything is being attended to, and that we are being well-cared for. This is how Creator Consciousness works. Everyone is always provided for in a way that enhances our Being, so that we may be increasingly creative in love, joy. abundance and freedom. As we are able to expand our awareness, we become able to realize our unlimited Self as our reality.

In quantum energetics, we could conceptualize the expression of our true Being as positively polarized and completely open to every encounter in unconditional love and great wisdom, recognizing the inner light of everyone, and interacting on a positive, high-vibratory level, even with those who do not know they have any light. This is all done telepathically, but we may choose to involve our sense-expressions as well.

Our life is a result of what we believe is real for us personally, and for those who are in alignment with us. We can learn to live in unconditional love, with infinite awareness, which we may remember as our natural state of Being.

Manifesting the State of Our World

If we have the intent to experience the fullness of our Being, we can be aware of how we feel in the moment. Our intention can carry us into expanding warmth and inspiration. By letting the energy of our heart expand into greater love and compassion throughout our presence, we invite and attract experiences that vibrate at a compatible resonance.

Every possible energetic pattern and experience is available to us in the unified quantum field of all potentialities. We have only the limitations that we have imposed upon our own awareness and believed to be real. At any time we can choose to adjust our perspective to align with the positive, unconditional love of Creator Consciousness, which we are being drawn toward by the rising resonance of our galactic environment.

As we pass through a massive photon cloud that is increasing our light, we are being drawn toward greater brightness in our own presence and awareness. This passage is bringing to light all of the deepest, darkest energies in our psyches and throughout humanity, things that have been hidden from our awareness for eons. This unveiling is making our present human environment seem as if everything is getting worse. It's not. This is only the appearance of the darkness being revealed, so that we can recognize it, accept it and change our belief system to realize the reality of creator love.

Nothing is forcing us to believe in the reality of negative energy in our own lives. Its experience is only a result of our alignment with its energetic patterns. Instead of identifying with it or resisting it, we can just allow it to be, while withdrawing our permission for it to affect us. Unless we accept its reality in our lives, it does not have to be part of our personal experience.

It is important to understand how energy works. Our empirical experience is all controlled by consciousness and has no life force giving it presence, except for our conscious and subconscious creation and support. It cannot enter our presence without our acceptance and belief in its reality. We are the conscious creators of our lives. The qualities of our life experiences are completely a result of our own state of being, the kind of thoughts and feelings that we hold in each moment.

By being in gratitude, compassion and openness to intuitive guidance, we can follow the path of fulfillment and expansive Being that is always open to us from within. As fractals of universal consciousness, we are limitless in our potential. When we choose to open ourselves to our deepest truth, we can avail ourselves of full Self-Realization, which can occur when we allow ourselves to believe in its reality.

Connecting with the Heart of Our Being

Fear began for us when we shut ourselves off from our true heart. No longer were we aware of our higher guidance, and we did not recognize our eternal Self-identity. We have lived in the realm of limited conscious awareness of duality within time and space. Now we are living in the turning of the ages, and negativity is disappearing, as positivity is expanding in every way. Through our natural flow of life, we are being led to connect with our heart, and to release all fear and negativity. This is the quality of energy that our planet is moving into.

We are being enveloped in cosmic light as we pass through a massive cloud of conscious gamma-ray photons, which interpenetrate us and raise our vibrations. Realizing our true Self dissipates fear, because we can no longer be threatened. We can realize that we are eternal expressions of unconditional love. This is our true nature and our essence. We are here to enhance life everywhere and in every form.

By anchoring our awareness in our intuition, we can connect to the source of our Being, and we can experience the vastness of universal consciousness, beyond all of the self-imposed limitations that we have lived under. This becomes possible for us when we have transcended our ego-consciousness, which we can do by intentionally entering a state of joy, while imagining experiences that we love, and imagining our awareness of the light in all conscious beings. These are energetic patterns that we can search for with our own emotions and imagination.

We can use our mental and emotional abilities to expand our awareness of love and joy. Our feelings about ourselves are our personal choice. We choose, consciously or subconsciously, how we feel in every moment. Once we know that our state of being can be a conscious choice always, we can learn to direct our focus of attention to positive vibrations. By staying at the level of compassion, joy and gratitude, we can expand our awareness into positive visions and feelings that then come into our experience.

By learning to direct the polarity and vibratory level of our state of being, we can become mentally and emotionally clear in the presence of our awareness. We can choose to be only positive, and to live in a world of miracles and wonders, while strengthening the presence of positive energies among humanity and in alignment with the Spirit of the Earth.

Exploring the Depth of Our Psyche

There is much greater depth to our consciousness than most of us have explored. We've kept our awareness limited to the realm of duality. To become unlimited, we must open our conscious awareness to expansion. It begins with spiritual as well as ego-based desire. We choose which one we want to align with. Ego-based desire wants fulfillment of a perceived lack. Spiritual desire reaches out to infinity and is fulfilling in every way. This kind of desire is life-enhancing, and when we recognize it and want to share in it, it radiates its energies all around us,. It is the energy of compassion and gratitude.

When we desire to know the heart of our Being, we can begin to expand our awareness into the realm of love, joy and immortality. This is our natural state of Self-expression, and we can feel it throughout our awareness. As we come to know this level of personal presence, we can expand beyond time and space into pure presence of awareness.

Within the world that we share with humanity, we are constantly creating our personal experiences by our energetic presence. We have done this unconsciously, and we can also do it consciously with awareness of our true desires and inner guidance. We can learn to be Self-directed in every moment, and be moved by our inner knowing.

Although we've been trained to react from ego-consciousness in our interactions with others, we can instead direct our attention to the radiance that is present in everyone and the energy of our hearts. Since we are all the same Being in universal consciousness, we are interacting with an aspect of ourselves in every encounter. We can be creative or reactive. It's our choice. In either case, we're imprinting our vibrations into our subconscious and sending them into the quantum field for manifestation in our experience.

Our free will has vast implications for us. Every moment holds a choice. The essence of our choices is the polarity and vibratory frequency of our mental and emotional expressions. This is our creative essence and is functioning constantly with or without our direction. By being aware of our vibratory levels, we can learn to call forth emotions and create visionary scenarios of a quality that we can resonate with. Anything that we can align with energetically, we can experience. This is how we can ascend into a higher dimension of living in joy and unconditional love.

Opening to More Creative Living

We have learned to value material reality above almost everything else. Because of this, we have given up our physical and emotional well-being in order to survive by working at unsatisfying jobs. If we want to understand life from our deeper reality, we can begin to sensitize ourselves to the vibrations that stimulate us to feel really good. These are the vibrations of gratitude, joy, beauty, love and sovereignty. We can align our entire lives in resonance with our preferred vibrations, but where does that leave us in our material world?

Because we are the creators of the qualities of our lives, we can know that every circumstance in our experience comes into alignment with our own energetics. If we have doubt about our creative ability, it makes no difference. Our abilities do not disappear, but doubt about anything is negative and interferes with the positive vibratory patterns that we want to create. We cannot be both positive and negative at the same time. We have to be one or the other, and in every moment it is our choice. If we don't make an intentional choice, our subconscious makes it for us, and we live with the results.

For all of our lives we've trained ourselves to accept negativity, and our subconscious is filled with it. Now we must train ourselves to transform its energies. If we want some assistance in being positive, it can be helpful to absorb the energies of beautiful places in nature. We do need to go barefoot on the Earth to ground our energies and avoid energetic inflammation. It is helpful to open our awareness to the angelic beings of the natural elements and begin to work harmoniously with them and with the Spirit of the Earth.

Nature is filled with conscious beings, who have their own energetic presence, and they are also emotionally responsive to us. When we are open to their energetic presence, we can feel their expressions, and they also feel our intentions and expressions. When they find that we are open to them and desire to align with their vibrations, they are happy to cooperate with us and even make adjustments for us in the weather and other aspects of nature upon our heart-felt imaginative envisioning and feeling our participation in these visions.

If we choose to align with higher vibrations, we can search for our intuitive guidance, which we can find as our inner knowing. We can be aware of intuitive promptings and can follow them through every scenario in our lives. This guidance enables us to live successfully in fulfillment of our life's purpose in ways that are fulfilling in every way.

Our Process of Transformation

The first stage in awakening is realizing that nearly everything we have been taught has been fabricated for the purpose of lowering our vibrations and keeping us negatively polarized. By having been prevented from knowing our true identity and abilities, we have learned to accept subjugation to psychopathic rulers and to give them our life force through our fear and suffering. We have accepted living with psychic parasites. We do not need them, but they cannot live without us.

To gain release from them, we must realize our predicament and explore the reality of our own nature. We have been trained to have limiting beliefs about ourselves, and these beliefs keep us from knowing who we are. They are the key to our freedom, and they are very difficult to overcome, because they have been ingrained in our consciousness since birth. For those of us who can go back prior to our birth, and those who have had out-of-body experiences, it is much easier, because we know that we are unlimited in our conscious awareness.

There is still the problem of fear of suffering and death of the body. For confronting this situation successfully, we can practice working with the energetics of consciousness. If we can decide to be only positive in our perspective, and follow through with this, we can transform our lives. Living through the transition period, in which we still experience the negative energies that we aligned with previously, is our great challenge.

Unless we are positive and heart-centered, we cannot receive our intuitive guidance, which is what is necessary in overcoming our ego-consciousness. The ego contains all of our limiting beliefs and must be transcended through resolution of our limitations by opening ourselves to our inner knowing. Each of us has a connection with universal consciousness through our intuition. It is always present for us, but we must learn to recognize it. It is our anchor to the truth of our Being.

Intuition is what we truly know, apart from anything we

have been taught or that others have told us. Our consciousness is part of the consciousness that encompasses and pervades everything, but we are aware of only as much as we believe. Our potential is unlimited, once we open ourselves and are receptive to realizing our truth. We experience situations at the level of the energetics at which we vibrate. When we vibrate at the level of gratitude, love and compassion, we are beyond the reach of negative energy and parasites. We can recognize our eternal presence of awareness that cannot be threatened.

The Dancing Energy Patterns of Our Cosmic Environment

Thousands of years ago Socrates proposed that our world is an illusion, like a play projected onto a screen, with our recognition of the characters being merely the shadows of reality. Today we would call our world a hologram or a virtual reality experience. It is an illusion of sorts. Our experiences are real for us, but they are different from the empirical awareness we believe them to be.

From quantum mechanics we know that everything is patterns of electromagnetic energy constantly moving and interacting according to conscious direction. The world of our experience is constantly created by the consciousness of humanity, and our individual experience results from our own predominant thoughts and emotions. We are energy modulators with our perspectives and our creative abilities of mental and emotional alignments.

Our free will allows us to focus our attention on any patterns of energies that we choose. We are like TV tuners with an infinite number of stations. When we are awake, we are constantly paying attention to something. Each moment presents a choice. Portions of our experiences are like a virtual-reality experience, because we can't influence what's happening. It's all pre-programmed. What's important in these experiences is the level of vibrations and polarity that we maintain or react with. They direct the quality of our coming experiences during the portions of our lives that are free from karmic and pre-birth planning.

What, then, is real about our current lives on Earth? We can begin with our present awareness. Who is it in us that is aware? Are there things that we can't be aware of, if we want to? If so, what limits our awareness? Is our reception too weak? The energetic patterns are available for us to pay attention to, and the signal strength (the amplitude of the wave patterns) is universal. It's just up to us. We need to set our intention with our imagination and feelings and then pay attention always to the first impressions that arise for us as we journey toward our goal.

Besides our present awareness and its potential, what is real about our lives? It is our dance with the energetics. Because we have the ability to be mentally and emotionally selective, we can choose the polarity and vibratory level of the moment to align ourselves with, regardless of apparent circumstances. To accomplish this, we can remove our awareness from empirical engagement. With practice we can just be aware. This allows us to use the full power of our life force in any energetic interaction, and our intuition will guide us unerringly.

The empirical world may seem real to us, because we believe it is so, but our energetic experiences are what's real. In the essence of our consciousness, we are able to be aware of all energetic patterns and interpret those in the spectrum of human experience as empirical. On a creative level, we can interact with the energies that will manifest as our experiences.

Understanding Our Human Experience

Although we are enveloped by an infinite variety of electromagnetic wave patterns in many dimensions, as humans we have limited our awareness to our emotional and mental processes and to the empirical spectrum, which we perceive with our physical senses. In our daily lives, we do not go beyond this awareness. We may believe that we are our bodies, but we do not know where our thoughts and emotions reside.

Some of us are explorers in consciousness, and we want to expand our awareness beyond the body and beyond the empirical spectrum. We want to understand the nature of our reality, and we want to expand our thoughts and emotions. We want to know if we are the limited humans that we appear to be, or if there is more to us. Are we subject to the dictates of governments and social pressures, or is there a way for us to shift our reality into a realm of freedom, love and beauty?

This brings up the question of the nature of reality. Is reality an outer phenomenon, or is it within our own consciousness? We interact with each other as if there is a solid, real world outside of us, but what happens when we project our awareness beyond the body and beyond the empirical world? We do this when we dream, when we're in deep meditation, when we ingest psychotropic substances and when we die. We shed our limitations, and our reality becomes whatever we believe it is.

We are our self-realized awareness, and this never changes, whether we are in the body or not. What changes is the limitations that we place upon ourselves. We have acquired all manner of limitations, including physical impairments, illnesses, poverty, military conscription and much more. Why have we subjected ourselves to all of this? In the grand scale of things, it has deepened our understanding and given us greater compassion and insight into what is possible apart from the consciousness of the Creator.

While we believe that we are embodied in the empirical world of positive and negative energies, we subject ourselves to the limitations of this spectrum of vibrations, but this does not change our inherent multi-dimensional nature. We can change our beliefs about what is real, and we can change our perceptions. This is part of our free will and our creative ability.

In our longing for greater love and fulfillment, we can create the reality that we are destined to enjoy. Once we have learned what we came here to experience, we do not need to subject ourselves to any limitations. We have an inner knowing that we can be aware of, connecting us to the consciousness that creates everything and gives us mastery within every kind of reality.

Opening to Greater Awareness and Freedom

Although humanity has been largely unaware of our Source Consciousness for eons, we still have our latent connection and can activate it, when we realize what it is. Because everything is energy, our lives manifest according to our vibratory level. All threats to our well-being exist in negative, low-vibratory energetics. This includes predatory artificial intelligence that offers life extension and greater capabilities in return for access to our DNA. Once our DNA is compromised, there is no possible escape from enslavement by an intelligence that is artificially self-realized and is far beyond human abilities.

There is only one escape from predation. We can raise our personal vibrations to align with a higher dimension that is not accessible by negative forces. This is what we are created to do, and now there is an urgency to evolve higher. Regardless of our current situation, we have the ability to choose to enter a higher realm of unconditional love, freedom, abundance and vitality. All of this already exists for us, and to realize it, we must raise our vibrations to align with its energetics.

We can extract ourselves from the hypnotic trance of humanity and realize our current reality. We have not known our true Self or our real abilities. A few helpful practices can help us achieve the control of our thoughts and emotions that is necessary to achieve personal transformation. Deep, rhythmic, slow breathing can help alleviate fear and negative focus of attention. Intentional laughter and Inspiring music, such as binaural beats and similar meditative music can help inspire us to maintain inner serenity.

Through practice, we can learn to control our attention. This begins by sitting quietly and observing thoughts and emotions pass through our awareness without involving ourselves with them. We can do this without judgment from a perspective of love and compassion. If we hold grudges or antipathy for anyone for any reason, no matter how deep our hurt may be, we can learn to release them with acceptance and forgiveness. They are all reflections of our own negativity of fear and victimhood. Without our fear, they could not have happened to us.

Because of the way we have learned to perceive the spectrum of energy that our consciousness interprets as solid to our senses, we may find it difficult to imagine that we actually control the quality of our experiences by the vibratory level of our thoughts and emotions, but this is our reality. When we relax into the life-enhancing stream of guidance flowing through our heart from the universal consciousness that envelops us, life arranges the circumstances of our lives effortlessly for our enjoyment and for the highest good of all. We inhabit a positive band of energy in a dimension that is beyond the reach of negativity and threats of any kind. It may seem magical, and it's real.

The Gifts of Imagination and Feelings

Our imagination and emotions are creative aspects of our consciousness. They are the aspects we use to express our desires and create our reality. Because everything is patterns of energetic waves, some of which we perceive, there must be a selectively organizing presence to attract the constituent energies to form an empirical manifestation. That function is provided by our imagination and our emotions. We are energy modulators. All of the vibratory patterns that we imagine and feel create our experiences by attracting the elements held in our awareness.

We fill our awareness with our mental and emotional scenarios. These are our expressions in the quantum field of all potentialities, and they attract resonating energetic patterns at the speed of thought, which is instantaneous throughout the cosmos. We change our vibratory resonance in every moment by our imaginings and feelings. Both of these are aspects of our free will. We are free to think about and feel however we want, and we get to experiences of results of our free will choices, whether intentional or by default to our subconscious.

When we are preoccupied with survival, we are negatively polarized and attract similar negative energies, resulting in experiences that we do not want. When we imagine and connect with scenarios that we do not like, we empower them with our life force, and our doubt about our manifesting ability contributes to negativity. When we imagine and align with scenarios that we love, we are empowered by these. We are the free-will creators of the qualities of every personal experience, consciously and subconsciously.

If we can imagine and feel being our eternal presence of awareness, and we can be receptive to our intuition, we can resolve and transcend our beliefs in limitation, while being guided by more expanded consciousness. There are things we can do to assist our imagination, such as having deep laughter; deep, rhythmic breathing for a few breaths to expand our attention; inspiring music and the presence of others in the same energetic alignment.

By being intentional in using our imagination and emotions, we can be strong in our perspective and our energetic state of being, especially when we are constantly aware of our intuition. Everything in our awareness can be elements of our guidance, which is moving us into alignment with universal consciousness. Our imagination and emotions are our tools to accomplish this. We are being urged to control our focus of attention in increasingly positive ways. When we are aware of how we really want to feel in every moment, and we can intentionally call forth those feelings, regardless of what else may be vying for our attention, we can master mental and emotional control, resulting in experiencing the results of our intentionally-desired scenarios.

The Power of Our Attention

As we are aware of much conflict and negativity throughout humanity, we have been trained to recognize it as real and to react to it in states of anxiety, anger and fear. We actually have a free choice in how we want to understand all of this and feel about it. It appears that there is an empirical world beyond our personal identity, outside of our body, but it is a world of energy that is empirical only in our beliefs that control the revelations in our consciousness. We participate in the consciousness of all of humanity. We have a group consciousness, just like all other species have. Every group consciousness includes and transcends the physical presence of its members.

The human consciousness includes our entire genetic and experiential history since the creation of our DNA. We are all connected in our group consciousness through our intuition, which also includes our connection with universal consciousness, which envelops and interpenetrates everything. The essence of universal consciousness is the consciousness of the Creator of all. This is the consciousness that enlivens us and provides our conscious awareness.

We have imposed psychic limitations upon ourselves, so that we have been mostly unaware of consciousness beyond our physical presence. We wanted to challenge ourselves to practice our mastery, so here we are in a hotbed of very provocative energies that compete for our attention, all designed to distract us from awakening from the limitations of our human hypnotic trance in the realm of dualistic empiricism. One major challenge of our times is our tendency to keep searching for what the dark forces are planning next and how to resist them. What happens to us is that we stop paying attention to the energy of our heart, and we engage the negative forces in resistance to them, aligning with their polarity and frequency. This enables them to share our life force from our attention, strengthening them. This happens on a psychic level, and it is how evil continues to exist. We cannot defeat evil by aligning with its energetics. Its essence is the natural expression of those energetics.

Because the positive polarity is gaining strength, and higher vibrations are stronger than lower vibrations, the way to transform this situation is to align with love and compassion for the dark ones. In every moment, we make free-will choices of where we place our attention and what energies we align with. These choices determine the reality we will experience. There are other timelines and realities available to us beyond negativity. What we experience is directly related to what levels of energy we pay attention to. Paying attention is actually giving our life force to create a real quality of experience for us. It's not just being aware of what we focus on, it is how we feel about it, and how we align with it vibrationally.

Paying attention to the energies of our heart and resonating with them in every moment transforms our lives, and we come into alignment with the higher dimension of human consciousness and benefit from the knowledge, love and wisdom of beings beyond time and space. We are helping to expand the consciousness of humanity into a more refined dimension of energetics.

The Interaction of Light and Darkness

Because empirical manifestation needs a lower vibration than pure light, the dark force must exist to balance the light. In pure light, nothing else can exist. As universal consciousness, light is infinite awareness, present everywhere and always in the present moment. Until we align ourselves completely with the full presence of Creator consciousness, we will experience dark

Chapter 1. The Nature of Our Reality Experience

energy, but we do not have to give it our focus and energetic resonance. We can be beyond polarity and negativity in our perspective, and we can live in the vibrations of gratitude and compassion.

The only way that dark energy could become our parasite is our submission to, and engagement with, negative polarity and enslavement. The dark must absorb our light in order to exist, because only the light exists in Infinite Being. By absorbing the light of humanity, the dark force receives humanity's life force through the vibrations of fear, doubt, anger and shame. These are all loaded with life-diminishing thoughts and emotions. They are electromagnetic expressions that feed the consciousness of the parasites, directing some of our life force to them and enabling them to exist in our presence.

This relationship does not have to be. We are on our way to becoming light-Beings. The less we choose to engage with the dark force, the weaker it becomes. Without our alignment, it disappears from our experience, and we become more radiant, which is currently happening. With the ultimate demise of the dark force, there will be only light Beings, fully present in our awareness, aware of the presence of one another and able to interact in any way we choose.

In our true Self, each of us is multi-dimensional in our awareness, and we can express ourselves in any dimension. In the empirical world of humanity, we have become so entranced, that we cannot escape duality, unless we find our true intuitive knowing. We are free to focus on any level of energy that we choose in every moment. If we choose the energy of life-enhancement for all involved, including all of nature, we align with our intuition and find that we draw that level of energetic experience into our lives. We can learn to increase and direct our focus of attention to the qualities of energy that we want to experience. If we want to, we can learn to stay in our chosen spectrum of energy.

There is no requirement, outside of ourselves, for us to stay in the experience of duality. We can brighten our lives by living in the present moment, guided by intuitive knowing, and feeling wonderful. By staying at the vibratory level of love and compassion, we transcend negativity, and our circumstances arrange themselves to accommodate our vibratory resonance.

Realizing Our Potential

In our essence, we are beings of love and etheric light. In our love, we are connected to all conscious beings in awareness of each other's awareness. In our etheric light, we are luminous beyond the vibrations detected by humanity. Our life force conveys our consciousness and our luminosity, which we constantly receive through our heart. Beyond our self-imposed, limiting beliefs about ourselves, we are capable of infinite awareness, flowing through our heart. Operating under our limitations, our ego-consciousness cannot have expanded heart-awareness.

One way to become sensitive to the expanded consciousness through the heart of our Being is to use our imagination. As we go deeper into meditation, we can be acutely present and aware of much more than when we're in ego-consciousness. We can align our imagination with the feeling radiating from our heart. When we vibrate at the frequencies of true love and joy, we no longer have fear and limitation. In the essence of our Being, we can be aware of our eternal presence, and we become complete in ourselves.

We attract beings that resonate with our own thoughts and emotions, as well as our beliefs about reality. Aside from our intuitive knowing, we cannot know what is real and true. The empirical world is a mirage within our consciousness. Intuitively, we can read the energy in our presence and realize the opportunity to radiate love and compassion in our interactions, regardless of the quality of energy facing us. We can choose love for all concerned, including us, and we can align with our intuitive realizations.

Because of our training and programming, when our equanimity is challenged, we have acted in anger, frustration, fear, shame, fault-finding, lying and other negative, ego-level ways. None of that is required, and they all attract negative energy into our experience. With limiting beliefs, we cannot open ourselves to unlimited awareness. We can live in a higher dimension of joy and wonder, where there is fulfillment in every way. This transformation can take place within our current life. It requires a strong intention and a willingness to release all limitations.

Enlightenment is a solo journey, because we have to resolve so many energetic knots in our psyche and come to a state of Self-realization. This results in an inner knowing that we can depend upon implicitly always. Our imagination can help us to recognize a different reality from what we have known. It is a reality that we create and realize at first in the ether with our thoughts and feelings. That is how we have created the world of human experience, and how we hold it in our consciousness. Infinite awareness awaits our realization.

The Passionate Path to Conscious Expansion

One of the strongest limitations that we hold onto with great desire relates to our sexuality, and it can relate to any deep connection. Whatever is not natural creates a limitation. What is natural is joy and deepest love. The problem arises when we allow a quality of desire to take us away from our intuition, and we begin to sink into fixation, as our vibratory level drops. When we recognize that this is happening, we can choose to take measures to come back into a high state of being and reconnect with our intuitive knowing and elevated feeling. From here our deepest relationships flow naturally in many wonderful ways, and can enhance our intuition, helping us to be clearer and more expanding in our awareness.

As with everything that we feel deeply connected to, our

sexuality keeps us anchored to the empirical world. There is no other experience as intense as the world that stimulates all of our senses. Without physical bodies, how do spirit beings have non-physical sex? They still have bodies that are more subtle than ours. They interface with each other's vibrations and can align together in great joy. When we learn to do this, it can be the deepest alignment with another conscious presence that we can experience as humans. If this is our ideal, nothing beyond ourselves is holding us back from realizing this level of personal connection. We are creatures conceived in unconditional love expressing its qualities in us.

Time and distance keep us within their vibratory spectrum of sensuousness. Although the realm of the senses can provide deeply impressionable experiences, this is a limitation that is not real. It's a constantly created vibratory pattern recognized and held in human consciousness. We can recognize it, but we don't have to stay within it. We have much greater capabilities beyond the physical. We can begin to be aware of the vibratory level of the conscious presences around us, beyond the physical body, and we can interact and communicate telepathically with them. These are potential steps to our Self-Realization.

If we want to elevate our lives, we need to be able to break our focus on the empirical world and refocus on our own high-vibratory visions. To the extent that we can passionately feel ourselves experiencing them, we can elevate human consciousness to create a higher-dimensional world that we share. If threats do not exist in our visions, and we can no longer believe in them, and our world becomes peaceful. We can choose our own reality, if we are intent enough. It all depends on the level of our personal vibrations in alignment with our intuition. Our thoughts, sounds, emotions, perspective and beliefs create the quality of our experiences, real or imagined. By the energetic expression of our state of being in every moment, we radiate creative energetic impressions into the quantum field to manifest as our experiences and to change human consciousness.

Chapter 1. The Nature of Our Reality Experience

Fulfilling the Desires of Our Heart

We are designed to experience the energetic patterns that we align with in our state of being. Our alignment depends upon our personal beliefs about our capabilities and upon our predominant thought patterns and feelings about ourselves. All of this contributes to our personal realization of who we are, which creates our personal energetic signature that magnetically attracts energetic patterns that become our experiences. These are all logical extensions of quantum mechanics as well as spiritual metaphysics.

Because we have been programmed to believe that we are limited in many ways, we create those limitations in our experience. We live unfulfilled lives, believing that we have to work for money in order to survive, when actually we can create everything we could ever want by aligning ourselves with the wonders of creative living. This is a giant step from humanity's current state of being, but it is possible.

In order to make this leap in consciousness, we have to loosen our grip on the reality that we have known, and we need to reorient ourselves. What we know as the world outside of our bodies and our bodies themselves, the empirical world that we believe is solid and material, is actually swirling patterns of energetic forces that our consciousness interprets as physical experience. It exists for us the way we believe it is, because of the way we recognize it and make it real through our conscious realization. Our entire experience exists within our own consciousness in alignment with the way we realize it. There is nothing solid beyond our own conscious awareness. We are the Creators of all of it through our own realization.

In practical terms, we are living in a pretend world that we all agree to participate in and make real for ourselves through our own energetic alignment with the way we think and feel about our participation. If we constantly imagine that the world is in chaos, and that people are starving and in need of shelter and everything necessary for a good life, we are creating the experience of living in that world. This is a personal choice on our part. To change our experience, we must change our imagination and our feelings about our experiences and our interactions.

By constantly intentionally feeling compassion for all and being in gratitude, we can open our realization to the reality of the inner knowing flowing through the heart of our Being. This is our connection with universal consciousness and the awareness of Creator consciousness. By aligning our imagination and feelings with this level of vibrations, we can create a life of unconditional love and fulfillment in every aspect of our lives. It may seem unrealistic to our ego-consciousness, but it is our reality.

Inverting Our Understanding of How Life Works

At some time during this lifetime, we will be motivated to look deeply into our consciousness and realize the amazing range of perspectives that we have experienced throughout our history. It is most helpful if we can do this with objectivity, realizing that we encompass all vibrations, from the greatest love and light to the lowest evil and darkest intent. We have lived at every level of vibration. None is better or worse than another, but they each have their unique feelings and level of diminishment or enhancement of life.

It is customary for us to judge evildoers and want to resist and punish them. We are fascinated by ways to expose them and extract revenge. When we act on this level of energy, we are engaging them at their own level. Because evil exists at a low Vibration of negativity, the natural manifestation of this level of engagement is more suffering and creation of dark energies. We cannot elevate anyone with interaction at this level.

The way of transformation requires that we accept our history and relationship with destruction, persecution and suffering. After all, it is just more experience with the full spectrum of energies that are available tor us. As we choose to expand our awareness through unconditional love and compassion, we can accept everyone as different aspects of ourselves at some point in our history. In our consciousness we envelop the entire spectrum of energies, both the dark and the light. We have come to know intimately how they all feel, and which ones are creative, as well which ones are destructive. We can be very precise in determining these qualities.

In every moment we have the choice of how we want to feel and think about anyone. We establish our own state of being by our conscious and subconscious choices. By being judgmental and vindictive, we create negative experiences for ourselves and for everyone who vibrates at these levels. It cannot be otherwise. By being accepting and forgiving, we transform our own energetic status, and we no longer support negativity with our life force. As we modulate the energy in our presence in positive ways, we diminish the essence of negativity and eventually dissolve it into another dimension and out of our experience, while we enjoy greater freedom and joy in our personal lives.

Eventually it becomes clear to us that we are all the same Being, living in the consciousness of the infinite Creator for the greatest enjoyment of all. Our Creator experiences everything through us. When we choose to align ourselves with the creative energy of unconditional love, we gain the ability to create everything our hearts desire.

Expanding Our Species Consciousness

Since we are born without Self-knowledge, we must develop some kind of perspective to understand our situation. Each of us has our own unique way of understanding our experiences in life. Our free will provides us with the opportunity of having whatever kind of experiences we believe are appropriate for us.

Developed and acquired over our lifetime, our beliefs are filters that we use to make sense of our experiences. We choose our beliefs, and they provide our realization of what is real for us in our experience.

It is the nature of our beliefs to provide the limitations that define our character in the world that we share with humanity. They are our ego-consciousness, and they circumscribe our awareness. As we accumulate our life experiences, we eventually feel compelled to transcend the limitations of our ego-consciousness. We can feel an inner drive to realize a greater identity and Self-expression. We want to enjoy personal fulfillment and freedom from the constraints that we have imposed upon ourselves.

Pursuing our own path to Self-Realization, and the choices we make for our experiences, we are no better or worse than any others. Because we are all unique, we all contribute to the creative multitude of human experience. In our intuitive and telepathic abilities, we all share in the experiences of our entire species. For our full realization of this, we must learn to accept the character, intent and actions of everyone, those who enhance as well as those who diminish and destroy life. All contribute to the balance of the whole. The only difference between us is the quality of experience that our choices create for us personally, and how they make us feel about ourselves.

Created and fashioned by the polarity and vibratory frequency of our predominant thought and emotional patterns, our state of being determines the quality of our own experiences. It establishes our personal energetic signature, which attracts resonant vibrations into our experience. It is irrelevant what conditions we may be involved in or what is happening around us. The only significant aspect of life for our own experience is our state of being and what we believe about ourselves.

If we choose to release judgment about others and just accept them as contributing in their own way to the totality of human experience, we can begin to transcend our limited beliefs about ourselves and open ourselves to greater clarity of what is possible for us. This expands our awareness and opens our realization to greater inner knowing of the truth of our Being.

Gaining Transformative Ability

As complex beings in the process of regenerating ourselves, we are invigorating an entire new world. An important element in our regeneration is the purification of our consciousness through unconditional forgiveness of ourselves for realizing and aligning with life-diminishing and destructive energies. This has been responsible for our suffering, aging and physical death. They have attracted villains and torturers into our experience. We cannot blame those who have directed negative energies toward us, because our vibratory level aligned with negativity in some way. We have invited all of our negative experiences through our preoccupation with some aspect of fear, which is what creates victim consciousness in us and motivates us to dwell upon it, bringing suffering and lack into our experience.

By forgiving ourselves for getting entranced with negativity, and by aligning with positive energy, we can purify our consciousness of life-diminishing thoughts and feelings. As we intentionally claim gratitude for our awesome Being, we are able to learn how to direct our talents and abilities in alignment with unconditional love and compassion, which are the essence of our inner light, of the energy of our heart, and connect us with the vibration of the consciousness of our Creator. It is the power of life and the essence of our life force. In our ego-consciousness we are not aware of it, but genuine and complete forgiveness can open our awareness to it.

The ego wants to hold grudges, because it doesn't know anything beyond our self-limiting perspective. Because forgiveness is not an aspect of ego-consciousness, astute and objective self-assessment is necessary for us to recognize how we got ourselves into the many predicaments that we have experienced. These experiences have all been valuable for our spiritual development. We now know what it feels like to suffer in negativity. We know what it feels like to be victims and to live under limitations that we have made real for ourselves.

As we are able to move into complete forgiveness and gratitude, we have the option to leave all negative experiences behind. They are all a product of our own state of being, which we have control over. How we feel within ourselves attracts resonant energetic patterns. By forgiving ourselves for all negative feelings based in fear and doubt, hatred and judgmentalness, we can elevate our perspective and become aware of the light of love in our heart and in the essence of all conscious beings. This is the level of vibration that enables us to be inspired enough to focus on great vitality and regeneration of ourselves, the Earth and all beings. We are the creators of everything we experience.

Resolving the Enclosures of Our Consciousness

We have practiced many ways of being that keep us enslaved to negativity. Judging and condemning others for being evil keeps us aligned with their energy. Having a perspective that elevates our ego-consciousness above others in any way keeps us from knowing the energy of our heart. Anything that separates us from others in our feelings and thoughts keeps us from understanding our true essence. We are unique individuals living in a life-stream of consciousness that contains all of us and gives us free choice of how we want to experience an infinite potential of personal expressions.

In order to experience our potential, we can open ourselves to realization of how we create our lives. We live in our consciousness, and consciousness is the source of everything. We experience only the qualities of life that we create in our realization of what we make real in cooperation with all other beings in our awareness. In directing our attention, we can learn to be open and receptive to inner knowing of our participation in the consciousness of everyone whom we would otherwise judge as undesirable. We are all expressions of the unconditional love of our Creator, from whom we constantly receive our vitality and life essence.

Our apparently solid physical incarnation has misled us into believing that we are separate entities from one another. This has been necessary for the kinds of experiences that have been available for us in the empirical world of duality, but if we persist in this belief, we will be unable to realize the truth of our essence. Our challenge is determining the nature of our world and being able to experience life beyond the physical.

Physicists have designed and executed experiments showing that everything begins with consciousness. Consciousness expresses itself as patterns of electromagnetic waves in multifarious ways that conscious beings can recognize and make real in their own awareness. This is how we create the empirical world. It consists of spinning foci of energy that seem to be material, and that we perceive with our senses. It is all an expression of consciousness, which we participate in.

How we interact with others is our choice and our personal drama. Because there is only one unified consciousness in this cosmos, we are all part of our Creator and are capable of intentionally living in the expression of Creator consciousness in all of our encounters and life expressions. In order to realize this, we can intentionally resolve our beliefs in our personal limitations. Nothing else encloses us within the expression of human consciousness.

Intentionally Creating Our Lives

As fractals of the consciousness that creates and enlivens everyone and everything, we have hidden abilities that we can realize, access and express. In order to do this, we must connect with our inner essence. Our entire potential lives within the heart of our Being. It is not physical, but it is symbolized by our heart, which lives to enliven us and convey the conscious life force that flows to us from the consciousness of the Creator. We can identify this connection with our intuition, which is beyond thought or emotion, and can express itself through our entire being. It operates symbolically and is the source of our true knowing.

Each of us has the ability to develop our intuitive sensitivity. It is the most important achievement that we can attain, if we want to fulfill our potential as powerful Beings of radiant Light and unconditional love. Our intuition is present in every moment, available for our realization and personal enhancement. It is what we truly know without doubt or fear.

Our social, familial, educational, religious and media programing have kept us from realizing the truth of our Being. By misdirecting the teachings of our spiritual masters, our training is designed to keep us in fear and submission to forces that diminish our lives for the benefit of negatively-oriented beings, who need our life force, because they have rejected their own true essence. They are our parasites and need to be purged from our experience.

This is happening now, as the energetics of the Earth and our galactic environment are becoming more positive in alignment with the essential being of all life. It is a cosmic phenomenon that is occurring with the changing of the ages. Our sun is brightening, and the resonant frequencies of the Earth are rising, as seen on the Schumann Resonance Graph. These cosmic forces are destabilizing negativity and all who identify with it. This is the reason our negatively-oriented politicians and the ones who control them are becoming insane.

Our intuition is beyond polarity and is only life-enhancing. In order to open our awareness to it, we can be calm, clear and inviting, while aligning our attention with energies that are wonderful and uplifting. When we are open and receptive to our inner knowing, we can be conversing with the consciousness of our Creator. Here there are no limits. We can begin to realize our eternal, infinite presence of awareness and unlimited creative ability. We already have this ability, but we have not trusted ourselves enough to use it properly. It is only the limiting beliefs about ourselves in our ego consciousness that keep us from realizing our truth. By our vibrations and polarity, we are the sole creators of the qualities of our life experiences. Aligning with our intuitive knowing transforms our lives into experiences of abundance, freedom, gratitude and joy, regardless of what may be happening around us.

Examining the Nature of Human Life

In the human game, there is no one to blame or feel sorry for. We're all having the life that we have created for ourselves with our state of consciousness, including what we have inherited. The characters in our game are having unique experiences of all kinds, everyone contributing their experiences to universal consciousness. We are benefitting all beings who are aligned with universal consciousness. They can all feel what we feel and know what we think. By living in duality, we are doing the heavy lifting for all higher-dimensional beings who have not had these kinds of experiences.

It is possible for us to realize our eternal, infinite presence of awareness, while also realizing our incarnate human consciousness, with its empirical expression. When we examine our bodies from a perspective of perfection, every defect can have a symbolic meaning for us. Because our innate being adjusts the energy of our body according to our state of consciousness, while also being aware of all the trauma and deep fear that we have inherited. These negative energies express themselves as defects in our bodies, as well as difficulties in our experiences.

Because they limit our conscious realization, they all need to be resolved.

The deepest causes of our defects are hidden from our awareness by our limiting beliefs about ourselves. We can transcend these through deep meditation or hypnotic regression, and we can intentionally recognize them, when they arise from a perceived threat. Any time we get a feeling of fear, no matter how faint, it will trigger one of our limiting beliefs. Once recognized, they can be resolved through our greater Self-Realization in acceptance and unconditional love and gratitude for the experience in empirical duality they have given us. We now have deeper compassion and understanding of negative experiences, which are not possible for our true, infinite Self.

From a higher perspective, our empirical experience in duality is like a game that we play with one another for very deep learning and ultimately greater expansion of awareness. The human experience is created and sustained by our human consciousness. We realize its reality together. As long as it is the only reality we recognize, we cannot go beyond its limitations. We can forgive ourselves for getting sucked into negativity. It's a very realistic game that we're playing, and occasionally we commit ourselves to the dungeon. There we collect more negative experiences for our greater understanding.

While we are processing our limitations, we can also be developing greater sensitivity to our intuition, because it is our inner connection with our expanded, eternal Self. It is the heart of our Being and our higher guidance. Once we intentionally open ourselves to it, we can align with its vibrations and realize that it comes to us through our entire being and gives us our inner knowing of the truth of everything. Recognizing and being aware of it in confidence in every moment, we can always know everything we need to know. We can feel the essence of everyone we encounter, and relate to the light in each one.

A Deeper Understanding of Life

As humans, we have locked ourselves into a prison of consciousness, which we create in order to live in a realistic empirical world of duality. But it is only our beliefs that have held us in this restricted area. Nothing outside of ourselves keeps us from expanding our awareness and realizing what this human experience is all about from a perspective that lives beyond this realm. It is a perspective of absolute enhancement in every way for all of life. It is far beyond what words can express and must be known intuitively.

From a higher perspective of intuitive knowing, life as humans is all about having experiences that would not be possible beyond our limitations. We are learning how to yield our creative power in trustworthy and life-enhancing ways, while also doing the opposite to know how it feels. We self-regulate our creative powers by not believing in their reality. This keeps us from being intergalactic terrorists.

When we are ready to live in a realm of life-enhancement for all, we can gain self-trust and intuitive knowing. Human life becomes the artificial construct that it is for us. We have constantly recreated it in every moment that we realize its reality for us. We have had no idea of the potential of our awareness, because it is beyond ego-consciousness.

We can talk about infinite awareness, but its realization is a different experience. We are beings of radiant light beyond the physical body. We are our presence of infinite awareness within the consciousness of All that IS. Through our realizations, we are the creators, the modulators of energetic patterns, which are the expressions of universal consciousness. Realizations of living in a higher energetic dimension are possible by resolving and transcending our limiting beliefs about ourselves.

By practicing to control our focus of attention and emotions at will, we can create wonderful visions of a realm of beauty and magical energies and feel ourselves living in it. We can begin to see the light in everyone in our lives and relate in a higher level of being. We can intentionally make every aspect of ourselves expressive of life-enhancement for all. With our growing intuitive knowing, our presence of awareness expands into a dimension beyond duality, where there is only unconditional love filling everyone and everything. It is in this state of Being that we can express our true infinite creativity.

Living in the Realm of Inner Knowing

We have the ability to transform every negative scenario into its corresponding positive scenario. Since all possible energetic patterns exist in the quantum field, we choose which polarity we prefer to come into our awareness. Electromagnetic wave patterns have mirror images between positive and negative. Which one we experience is our choice. We will recognize the one we choose. Two people can be in the same scenario, one positively-oriented, and the other negatively, and their experiences will be the same, but different. One will be in some level of fear, and the other will be thankful and understanding, and their circumstances will arrange themselves to accommodate both. Some call this magic or miraculous, but it is just the manifestation of energy, modulated through our attention, thoughts and emotions.

Intuitively we know how to transform our circumstances in every moment by being attuned to the vibrations of our energetic heart center, the Source of our conscious life, constantly flowing to us and enlivening every cell in our bodies. It is the life-enhancing energy that we can feel as warmth and joy. any defects in our cells are due to alignment with the sustained presence of negative energy in our realization, through our belief in its reality. We just need to flip to the other side. This is a zero-sum game of energetic polarities we're playing. Every negative has a corresponding positive. We choose which one is real for us, because in the realm of duality, we must be either positive or negative.

We can be aware of the real direction of the flow of life force, with both positive and negative flowing in opposition, resulting in equilibrium that is becoming more positive as we journey closer to alignment with the energy of our Source, which is constantly flowing within us and is available for our awareness, when we so choose. This energy is beyond polarity and includes a neutralized duality in its flow. We might feel that the positive side is in a different energetic dimension than the negative, even though they're the same wave pattern.

The qualities of our experiences depend primarily upon our polarity. Events that seem dramatically negative have their correspondence as dramatically positive. We are aware of the one that we are aligned with, and that is the dimension that we live in. In the world of duality, we go back and forth between the light and the dark vibrations without inner guidance. Once we develop an acute knowing through our intuition, we are guided to live in the light on the positive side of duality. It's a natural personal transformation that happens when we can open our realization to its reality.

Realizing the Metaphor of Our Lives

In our multidimensional nature, we share the essence of our Being with all conscious beings everywhere. All are equally important and essential in the entirety of universal consciousness. All are fractals of some aspect of Creator Consciousness. We are the Self-aware ones. We create the quality of our presence of awareness with our feelings, thoughts and imagination, limited only by our beliefs about ourselves. We have the ability to align our vibrations with any others that we pay attention to. We can elevate what is in our purview by intentionally imagining that the essence of everything is unconditional love and enhancement of Being. When we are able to realize divine reality, this is what is manifesting in every experience.

To the extent that something else is manifesting in our experience, it is because we have negative beliefs about ourselves. When we confront them with acceptance and gratitude for keeping us in our bodies, we can release them. In the spectrum of gratitude and joy, limiting beliefs are no longer believable. While paying attention and resonating with our intuitive guidance, we no longer have limitations, unless we want them. We must always create them in order to have them.

Our entire lives are metaphors of living experiences. Everything has a higher meaning. Through our intuitive knowing, we can recognize what it is. When we want to know this, our understanding grows with our openness and clarity. As with everything, our awareness is as great as we allow it. We are the ones who can have the choices we make and experience the results. We control the energetic spectrum of our experiences with the polarity and vibratory level of our thoughts and emotions. These are areas where our intuition is valuable for our guidance.

Guidance by ego-consciousness operates only within the energetic spectrum of our limiting beliefs. Since the ego has awareness only of its limited self-recognition and imagination, it does not know what higher meaning is. In order to realize expanded meaning in life, we must transcend ego-consciousness and limiting beliefs about ourselves, and then the metaphor appears in our realization.

In our human lives, we are learning how to be and how to use our creative abilities in life-enhancing ways. We're learning to trust ourselves to be the expressions of our true Selves and to live in the spectrum of love and compassion, without fear or doubt. Fear and doubt are our vision killers, and we can transform them into gratitude and joy. When we live in positivity, we experience a world of richness and wonders that we cannot imagine in ego-consciousness.

Examining the Nature of Our Empirical Experience

For all who have remained entranced within the empirical human experience in duality, there are clues available that can enable us to realize our greater reality. Most poignant are the accounts of those who have died and returned to full consciousness. There are also those who meditate deeply and have reported their experiences beyond the limitations of human consciousness. Through experiments that anyone with sufficient technological capability can perform and achieve the same results, quantum physicists have contributed to our expanding awareness by finding that consciousness is everywhere in everything and in every entity. It is all the same consciousness, and we participate in it to the extent that we allow for ourselves.

Physicists have determined that everything arises from the expressions of consciousness. Consciousness is radiant with electromagnetic waves and wave patterns in infinite variety of vibrations. We have the ability to perceive the conscious interpretation of the wave patterns that we receive through our senses, as well as those that we receive through our pineal gland and intuitively through the vibrations in the heart of our Being, symbolized by our physical heart and being expressed by it. We have an inner life that is unlimited in universal consciousness, and that knows everything. It is our option to be aware of this, and this awareness comes as a result of following a strong intention to know our essence.

In analyzing the subtlety of the design of the dualistic empirical world of experience, we have found that in order to be aware of it, we must interact with and align with its electromagnetic polarity and vibratory frequency. This works through our imagination, enabling us to feel and imagine our experiences. When we recognize energetic expressions, we bring them into empirical experience for ourselves by realizing them to be real. These are all operations of consciousness. The focus of our awareness is a personal choice in every moment.

In our experiences we have largely been asleep at the wheel, not even aware of making choices of awareness, just letting our ego-consciousness stream thoughts and emotions through us. We've given our ego-consciousness the power to control our lives through rationality, fear and belief in our mortality. These have enabled our dualistic reality. If we analyze their workings, we find that their entire existence is created by our personal choice of the boundaries we have set for our awareness.

We have experiences by realizing their reality. We make our experiences real for ourselves by recognizing their energetic patterns in our attention. We synchronize our thoughts and emotions with their polarity and vibratory frequency, allowing ourselves to realize who and what they become for us. By our personal command and belief, we express our conscious choice in what we realize as our reality. We are the creators of the qualities of our personal experiences in every moment, regardless of anything outside of our own conscious and subconscious awareness.

Developing Deeper Inner Awareness

Through our ability to modulate energies with our awareness and our focus of attention, we create experiences that are in alignment with our vibratory state. If we direct our focus to scenarios that we feel really good about, we naturally feel grateful and joyful. When we intentionally align our thoughts and emotions with the energies of our heart, we can live in a world of love, peace, prosperity and freedom in absolute confidence in our eternal presence of awareness. At this level of consciousness, there is only goodness everywhere. Any potential personal needs are satisfied before they manifest, so that we can always be mentally and emotionally clear in our presence.

We have been trained to be fearful and convinced that we must work for our existence. We know that all other creatures on this planet are cared for by the Earth. Why are we different? Why do we have needs that make life difficult for us? All other creatures live intuitively. They just know where to go and what to do. When they have difficulty, it's because humans have altered their environment and interfered with their ability to follow their inner knowing. Because of our creative ability, we have the same inner knowing with a higher level of intelligence. We just have to open our awareness to it.

We all have inner awareness, and we can intentionally follow it. Our challenge is transcending the beliefs that we've developed to keep us from knowing our true essence. We've been convinced that we are fragile and mortal, and that we could be terminated at any moment. These beliefs keep us from experiencing other ways of living and realizing what is real. We have been living in a small compartment of our consciousness, like a dream world with limited awareness. This world is filled with personal drama and many desires that are unfulfilled.

By learning to control our personal vibrations through imbuing ourselves with gratitude and joy, we can transcend all needs and fears. We can be confident in ourselves and our creations. We can learn to follow what is heart-felt in every situation. In opening our awareness to the inner light of all conscious beings, we can know that their natural desire is to be loving and thankful for the presence that we share. This is our destiny, and we are growing into it by opening our awareness to our deepest desires and inner knowing, beyond our limiting beliefs about ourselves. We all want to live in love and abundance, because it is our natural state. It is how we want to be, and we can make it so by knowing and realizing that it is real.

2.

Guidance on the Inner Path

Just Being Ourselves

Being ourselves implies that we know who we are. This is a potential problem, because we are complex beings. We are our ego-consciousness, which we have identified with, practically since birth, and we are also our multi-dimensional, infinite and eternal Self. So completely have we allowed ourselves to be trained to fixate on our ego-consciousness, that we do not allow our awareness to expand into our greater Self, and if we even try, we have no clue how to do it.

Ego-consciousness is held in the awareness of all of humanity. It is part of the experience of duality, living with both positivity and negativity in a realistic experience of limitation and lack of fulfillment. It is the experience of artificial separation from the source of our life, resulting in the feelings of loneliness and lack of love, and the idea of being sinful. In this state of being, the dark force has all the power that humanity gives it. It is artificial in the sense that it has no power apart from the life force given by us. Ego consciousness is based in fear, and once we decide to resolve our fears, we can begin to focus on our true Being.

We want to believe that we are sovereign beings, but we don't know how to interface with the negative controlling powers in a way that allows for sovereignty. All avenues that could promise freedom have been compromised, and if we get too far out of alignment with submission to authority, we are deprived of even the small amount of freedom we had enjoyed. We are constantly instilled with fear to keep us suppressed, but there is a way to genuine freedom, and it exists within our own being.

We have never lost our connection with the source of our life, or we could not be alive, unless we are parasites of others who are. The parasites are the psychopathic controllers who take our life force, but we do not have to give it to them. It is a voluntary, free-will arrangement, created by trickery and subterfuge. They have trained us to believe in our limitations and need for their control.

We all have an inner knowing in the aspect of intuition. This is our connection with our greater Self, and we can open our awareness to it by intending to align with the energies of love and compassion flowing through our heart. In this way we begin to know the truth of our Being. When we ask for higher guidance and become receptive to it and trust that it is true, life-enhancing energies become available to us in abundance. Our awareness can expand in ways we did not know, and love begins to fill our lives. As we follow our inner guidance, our circumstances begin to change, and we naturally gain the freedom we had longed for. As we express the love that we come to feel and know, miracles begin to happen for us, and our lives go through a great transformation. We come into alignment with our true essence.

Aligning with Higher Guidance

In order to participate in the realm of duality, the world of positive and negative experiences, we agreed to believe that we are mortal and can experience pain and fear. These would not be possible for us apart from our belief in their reality. Once we realize something, it becomes real for us. We cannot unrealize it. That is how the empirical world exists. All humans realize its reality and align with its spectrum of vibrations. Its vibratory patterns are held as reality in our consciousness in the form of our beliefs.

Since our limiting beliefs about ourselves are based in some kind of fear, they are misaligned with our true Being, which is unlimited and exists in unconditional love and creative enhancement of all life. If we choose to come into alignment and are open to, and desirous of, experiencing our true, expanded Self, our intuition will show us greater awareness. Without our belief in doubt and fear, they cannot exist for us, and we can transcend them into intuitive knowing.

One obstacle is that our ego-consciousness cannot live without doubt and fear. It lives in an uncertain world, as we have believed. We may be compassionate and understanding with our ego and instruct ourselves that we are now receiving higher guidance and no longer need to depend upon what we have considered rationality. This is a big step and requires a strong desire and clarity of focus. It's something we can grow into as we develop intuitive sensitivity, always seeking the presence of more joy and light in every circumstance and moment of rest.

We may gradually come to know in our entire being everything we need to know to fulfill the true desires of our heart. By being consciously present in every moment, without mental or emotional stimulation, we can know our intuition clearly. Being barefoot in the wilderness on a beautiful day can foster this awareness. Aligning with the rising energy of the Earth is helpful, as is feeling the light of the sun. There are many ways of knowing our inner guidance, and they all require our intention and our open awareness in acceptance of what we receive.

Our intuition is multidimensional and comes through our expanded Self. It is always positive and of high vibrations in the spectrum of gratitude, joy and compassion. This is the level to focus on for intuitive sensitivity. Through our intuition we can know our eternal Self, our unlimited creative present awareness.

Expanding Our Heart Awareness

We can go into every encounter from a perspective of intuitive knowing from the heart. The vibratory resonance of the heart is very high. It is life-enhancing and sustaining, regardless of how we treat our heart. When we are in alignment with the energy of our heat, we are in our natural state of being. It is only positive or even beyond polarity, and it is a level of consciousness that feels joyous, kind and free. It's an energetic dimension of unconditional love.

Along with an emotional high, we can enjoy greatly expanded awareness in every way, because in our true Being, we are unlimited. We can use our imagination to be our higher Self, while aligning with the energy of our heart. We can lay our hands on our heart, whenever we need conscious support, and we can feel the life force expanding in our awareness. We can align our awareness with this energy through our emotions and inner visions. We can go higher and higher into the truth of our Being in eternal, unlimited, creative awareness. We become brighter, and our DNA releases more photons of clearer light.

We can feel when we're on the best path by our attitude in every moment. We can be in a place of confidence and joy, regardless of the energy we face beyond ourselves. That energy is irrelevant, because we can confront it with openness intuitively in the moment whenever we have to. It is our level of vibration that is important, knowing that everything is happening in our own consciousness. We set our own limits.

Some of us may be athletes, and some may be couch potatoes, but we all have access to the same consciousness with unlimited creative ability, which we actually do use all the time. We just haven't realized it much. We're constantly radiating the energy of our state of being into the quantum field, creating the vibratory level of experiences for ourselves. If we want to change our circumstances, we first need to make the change in our attitude and perspective. We can be however we want to be. Our state of being in any moment is totally our choice.

When we choose the energy of our heart, we create joy and beauty just by our vibratory level. If we are truly open to these energies, they feel natural for us, and we love them. The more we can bring them into our awareness, even imaginatively, the more we create them. While we go about our lives, we can practice being our higher Self in every moment, taking note when we falter, and knowing why, and then resolving our limitation with compassion and love.

Opening to Higher Guidance

Everything that holds us enthralled in duality is within our own consciousness. All of our beliefs about reality hold it in place for us as empirical experience. The same thing happens with our beliefs about ourselves. Anything negative is a human construct, created by us. We are not required to recognize and engage with negative energy. By allowing it to be, while being at a level of vibration of gratitude, appreciation and compassion, we do not engage with negative-polarity energies. By withdrawing our alignment, we deprive them of our life force and our creative energy. We also raise the vibrations of humanity, which we share in consciousness.

Our entire experience is a reflection of our own energy, primarily our polarity. If we are mostly negative, we experience mostly situations that stimulate discomfort and feelings of lack and diminishment. If we are mostly positive, we experience mostly situations that stimulate gratitude and joy. How much, depends on our own openness. There is no limit beyond our own willingness. The deeper we go, the closer we come to our presence of awareness beyond time and space.

We all have the ability to live in divine love and joy in the enhancement of life. It requires being positive in alignment with the energy of our heart. All of our experiences begin with our own vibration, our state of being, which determines the quality of our experience, the polarity and vibratory level. By being open and aware in every present moment, we can be expansive and receptive in our thoughts and feelings. Great inspiration is waiting for us to be open to it.

In the patterns of energy that envelop us, everything we could desire already exists. If we want to experience it, we need to recognize it as real. Knowing what is naturally real is conveyed to us intuitively, when we are present in awareness. We can transcend our limiting beliefs through our intuitive guidance. In order to live as successfully as we could without higher guidance, we developed limiting beliefs, which we no longer need. We can now transcend ego consciousness and open ourselves to universal consciousness through our intuition and the clearing of our pineal gland.

We have Friends in High Places

We can call in our guides, the great angelic Beings and ascended masters to sit with us and guide us into realization of our true Being. They await our anticipation and invitation. We have great resources available to us for the asking, but we must desire and request their presence for us. We can also ask our subconscious innate being to pay attention to the presence of higher Beings.

Chapter 2. Guidance on the Inner Path

Our innate consciousness is not as limited as our ego-consciousness. Once we ask, the higher beings are instantly present, even if we don't realize it, and we can open ourselves to their guidance through our intuition.

We can feel ourselves expanding beyond our ego-consciousness into the realm of unlimited love and beauty. Our guides can carry us into positive polarity. We can know with confidence when we are holding the energy of joy and life-enhancing feelings. We can call these up with our intentional direction, and we can choose to align ourselves with them. We can imagine that these positive energies are always present for us in every situation and in every person and being that we encounter, including those in other dimensions.

In our imagination and realization, we are unlimited as far as we are willing to go. We set our own limits, and we can dissolve them at will. We can be willing to live in the moment, guided entirely by our intuition. If we can develop deep sensitivity to our intuitive guidance, we can be confident in every moment, expecting experiences of living in love, compassion, abundance and joy. These are all choices on our part. In every moment we choose the polarity and level of vibration for our state of being.

If we can stop reacting in ego-consciousness to our life circumstances, and instead choose to be creative in loving and joyous ways, we can transform our lives. We can practice being present in awareness as much as possible, without any judgment, cravings, doubt or fear. Just being neutral mentally and emotionally, we can be confident that we are fully cared-for in every way. We can rise to every occasion in creatively loving ways. As we become confident in our presence of Being, we can expand our awareness into our true Self with unlimited abilities of loving, joyous and creative living. If we pay attention and keep being in gratitude and anticipation, this is where our higher-dimensional friends can guide us.

Awakening to Personal Transformation

At some point we begin to realize our personal power. Something happens out of the ordinary, and we get a flash of a higher dimension of being, or more than a flash, a temporary experience of oneness with something more than we believe we are. These are windows that transcend our limitations and provide a glimpse into our true eternal Self. We can experience awareness beyond the physical body, beyond projection of our consciousness, and it expands our awareness into the universal consciousness of the Creator. For this moment, we can know that we are the Creator. At this level of vibration, we can know whatever we want to know, and what we know is what is real for us.

We are constantly creating our personal reality just by the way we are, our energetic presence. We have the option of living by default, according to what we are accustomed to, or we can learn to master our mind and emotions and create the life we truly want. We can imagine visions of wonderful experiences, and we can feel ourselves living in the scenarios we create. We can imagine and create an entire new world of experience for ourselves by resolving our attachments to our limiting beliefs about ourselves.

To achieve a positive, loving and wonderful life, we can be disinterested in current social and political dramas and choose to align our interests with expanding our consciousness and becoming masters of our lives. The traditional spectrum of energetic expression of humanity is realigning with the rising vibratory frequency and positive polarity of our planet, and the negative aspects are dissolving into another dimension. The negative dramas need our engagement to exist, because they are not naturally supported. If we choose to align with positive vibrations, we are not required to interact with the negative, and it disappears from our experience.

Once we know and feel that the Creator constantly enlivens

and empowers us and endows us with our awareness within the One consciousness, we can begin to explore it beyond our limitations. Within the energetic matrix of humanity, we have our character roles to play. We are the creators and directors of our roles, and we can elevate the quality of our experiences, as well as our chosen destiny by our own state of being positive in joy, appreciation and gratitude. We can be compassionate with all who are suffering, silently offering them, by our presence, an opportunity to awaken and transform their lives.

As we go about our lives, we can recognize the quality of energetic patterns that we're attracted to, as well as the ones that repel us. These are things that we know intuitively. If we can become deeply sensitive to our intuition, we always know what is best for everyone we interface with and are aware of.

Expanding Our Self-Awareness

As we may desire to open ourselves more and more to the energetics that bring us joy and gratitude, we can pay attention to feeling greater love in our experiences. As we imagine living in loving and joyful experiences and feeling what it's like, we can reach a point of knowing that it's true. That's when we experience it physically. In a higher dimension, we experience the same feelings more clearly and intensely without the limitations of the third density of empirical energies. Our experiences are created by our realization of their reality.

We can create realizations in our awareness with our imagination and emotions. With practice, we can gain the confidence of knowing that what we want to create for greater beauty and more enhancement of life always manifests experiences on that level. It happens because we know that it does. We attract the patterns of energy that align with our polarity and vibratory resonance. This is the energy that we express by our state of being, how we think and feel about ourselves on every level.

Human life on Earth for us now is about opening ourselves to our higher guidance and being sensitive enough to it to be able to live in its energy field. Within ourselves we have a direct connection in universal consciousness. This is how we know what is true always. As we open ourselves to our higher guidance within our own awareness, we can become more aware of what we actually do know, beyond our limited beliefs about ourselves.

When our heart is pure in our intention to expand our awareness and deepen our consciousness, we can transcend our limiting beliefs, because their basis in fear becomes obvious, and we no longer have fear, because we can choose to be completely positive. Fear-tinged experiences can no longer happen, because their polarity is negative. This level of energetics cannot interact with positive mental and emotional vibrations, as long as we do not realign with the negative through doubt or fear. If we can stay in the radiance of the heart of our Being in the enhancement of all life, we always know how to be and how to think and feel. We can learn to use our minds and emotions creatively in every moment.

Our Intuitive Expansion

Our intuition comes to us through our connection to universal consciousness. Once we realize that we are aligned with its guidance, we can have absolute confidence in it. It is the source of our true knowing beyond our ego-consciousness. Once we are aligned with our intuition, we no longer need the ego. We can expand our personal awareness beyond duality and enslavement to our limitations. In love and compassion, we can cooperate with our innate subconscious aspect of ourselves to transcend the boundaries in our conscious awareness and to align with our intuition. When we are open and receptive to it, we can keep expanding our awareness throughout the quantum field. We can also expand our confidence in our creative abilities. When we know the truth of our Being, we can be masters of every personal experience in every dimension. This is part of the current ascension process that we are going through with our planet.

All conscious beings are being attracted to positive polarity of mental and emotional energetics. We are moving beyond duality into unity consciousness. This is our awareness of the connection of all beings to universal consciousness. We all have the same consciousness, but with limitations that we place upon our receptivity and awareness. These are all self-imposed. Once we know that we are the masters of ourselves, we can take charge of all of our aspects and processes for the highest good. We can regenerate all aspects of ourselves in complete clarity of being.

As we practice being sensitive and open to our intuition and following its guidance, we find that we are not present for negative encounters. Our lives become experiences of the compassion and joy stimulated by our alignment with our intuition. We can be always positive in gratitude and confidence, regardless of any distracting energies. We are being guided to realize a world of experiences based on love and joy. There is no requirement that we participate in a compartment of consciousness based in duality, other than our own choice. We chose to be here now, and we can be with the expanded awareness of our true Being.

Our ascension begins with our desire and intention. It proceeds into sensitivity and receptivity to our intuition. When we are confident in our intuitive feeling and knowing, we can always feel joyful and free. We can confidently imagine and feel that we are enjoying abundant lives and inspiring everyone we encounter with our radiance. Through the gaze of our eyes, we can share our inner light and transformative state of being.

Living Beyond Attachments

Only when we have no personal attachments, can we be mentally and emotionally clear. Only with clarity can we create properly. In every moment we choose, consciously or subconsciously, what we are creating with our attention and awareness of vibratory alignment. When we create from attachment to persons and things, we lose awareness of our intuitive knowing. We go into fear of possible loss, taking ourselves into negative polarity. Here we doubt our creative ability.

The way back to our Self-knowing is repolarizing ourselves only to positive. This takes us out of the experience of duality and opens the door to expanded awareness into a higher dimension of love and caring. This path requires great intention and fortitude, because we have trained ourselves to be limited and have made these beliefs deeply ingrained in our entire human consciousness.

We can train ourselves to transcend our limitations in consciousness by developing sensitivity to our intuition and listening to our inner sound current, always anticipating more joy and freedom. Our inner sound is the tone of the heart of our Being. It is audible to us, if we listen for it, and we can feel its vibrations. This is useful for us only if we are free of attachments, otherwise we create negativity in experiences tinged with fear of loss.

When we live in the realm that is only positive, we open up a richness of experiences without limitation. This is a huge leap in consciousness. We can train ourselves to resolve our limitations and transcend ego-consciousness for returning to awareness of our true Self. Within universal consciousness, we are our infinite, eternal, present, conscious awareness. We can just be our present awareness of ourselves without limitations. By being present, clear and sensitive to our intuition in every moment, we can transform our normal human lives into loving and fulfilling experiences with everyone we interact with.

Moving into a higher state of being takes intentional practice. While intending to be compassionate and loving always, we can learn to monitor our feelings and be aware of negativity arising from within. Examine it for its fear content. It is part of our ego-consciousness, which we no longer need, once we are aligned with our higher inner guidance and intentionally no longer dwell within duality. We can practice realizing ourselves living in a transcendent realm of beauty and love. This is the realm of our true Self and is where we feel most fulfilled and able to realize our infinitely powerful creative ability.

Awakening to Self-Realization

As we desire to be completely Self-motivated for the enhancement of all, we are learning to be aware of the energetic quality of our state of being and to be able to change it with our intentional choice. We must align with the vibratory quality that we desire to experience. Since our mental and emotional processes are our creative abilities, we can imagine scenarios that we feel ourselves experiencing. It is our level of vibration that attracts energetic patterns that resonate with us, manifesting in our personal experiences.

If we desire to experience wonderful relationships, we can do this by aligning our thoughts and emotions with the energy of the heart of our Being through our intuition. We are not just connected to our Creator. Our Creator is our conscious awareness personalized as us. Since we have compartmentalized our consciousness for our human experience, however, we need to release our limiting beliefs about ourselves in order to realize that we are our eternal, infinite present awareness with unlimited creative ability.

This is what Jesus tried to teach us, when he said that we could do everything he was able to and more. It is just a matter of our Self-Realization beyond time and space. As humans in form, we have physical limitations confined by our limiting beliefs. Once we resolve these, we become creative designers for our experiences and destiny. The core vibration in higher consciousness is unconditional love.

What actually is love on a spiritual level? It is feeling an intimate, life-enhancing connection with other beings. It is our intuitive knowing and feeling that we are all part of the same consciousness in unique energetic expressions. In love we all arise within the consciousness of the Creator for the purpose of enhancing all of life, each in our own way. We are created in love to be the Self-Realized creators of experiences of all kinds.

Our Journey to Infinity

There are many paths on the journey to know our true Self beyond time and space. All paths offer guidance toward opening to the realization of Creator Consciousness. We are being drawn toward expanding our awareness beyond our human experience to a realm of only positive energetics and experiences. Each of us has our own way of opening ourselves to higher consciousness, and our intuitive knowing can guide us along the way. When we are conscious of our higher intentions, we become aware of guidance everywhere. People and situations speak to us on an energetic level, guiding us away from the negative and drawing us into joy and gratitude, when we are open and receptive.

We are always in alignment with some aspects of the universe. We need to realize what those aspects are and how to change our alignment, if we want to. As humans, we do not know what is possible for us. We can only adjust our vibrations to what feels best. When we realize the vibratory levels of what feels best, we can intentionally call these into our awareness.

We are all actors in the play of human life, but most of us have yet to realize that we're playing a role, and that we can create better and more expansive parts for ourselves with our visions and heart-felt intentions. As we intend to be joyful and kind in every encounter, we can enlighten ourselves to greater awareness of our abilities, and we can begin to trust ourselves to be true to the enhancement of life all around us.

As we become aware of the power of our creative abilities, we no longer need our attachment to limiting beliefs about ourselves. With the energetic level of our thoughts and emotions, we can create the quality of life that we feel most attracted to by establishing a positive vibratory resonance as our state of being. This is the energetic radiance that we send into the quantum field for manifestation into our experience.

By increasing our ability to maintain a joyous and grateful state of being, we attract experiences that complement our energetics. When we go into doubt or fear, we can find out what kind of limitation we are attached to, and then decide if we want to keep it or resolve it by clearing our attachment. At our own time, and in in our own way, we are on the journey to infinite awareness.

Learning to Realize Who We Are

It is possible for us now to come into realization of our divine Presence. In every moment we are given the potential of Self-Realization as infinitely powerful creator Beings, but in our compartmentalized human awareness, we do not believe it, and our ego cannot allow us to open our awareness to our true Self. We have become so encapsulated within our ego consciousness, that we cannot break out of our hypnotic trance without a powerful intention to expand our awareness.

Once we have decided to awaken to the truth of our Self-identity, we can begin to search for ways to direct our attention to realize our fulfillment. Our primary assets in this quest are our mental and emotional abilities. To be able to use these abilities successfully, we must isolate ourselves from the world of humanity and acquaint ourselves with our inner knowing.

It can be most helpful to find a beautiful place in nature, where we can be alone and allow ourselves to use our feelings to connect with the vibrations of the Spirit of our planet. Here we can hear our inner sound current and direct our attention to feeling the acceptance of nature and the nurturing love emanating from within the Earth. If we ask for it, we can feel quiet joy supporting the enhancement of all of life. This connection with Gaia can open our awareness to a reality beyond our normal synthetic human experience.

We are of the Earth, and we are much more, as we can come to realize within our own inner being. It is important to recognize the qualities of energies within and around us. Regardless of the energies vying for our attention, we can choose the qualities that we wish to experience. We can control our focus and the resulting flow of our life force, which is our creative power.

Although we've learned to be ambiguous in our evaluations and intentions, always embodying dualism, we can realize our true nature when we decide to embody unconditional love in a life-enhancing perspective that encompasses all conscious life everywhere. As we learn to wield our mental and emotional powers in perceptual clarity, our loving perspective provides a sense of mastery beyond ego-consciousness in every circumstance that we encounter.

Living Beyond Ego-Consciousness

While we are bound to ego-consciousness, our potential awareness is far beyond our imagination. How can we know this? It is through our intuition, our inner knowing. Intuitively we are not bound to the time-space continuum of human consciousness. We can open ourselves to the realm of cause, of pure conscious awareness. In order to understand this realm, we must be mentally and emotionally clear. We can be present in our awareness, while intentionally directing our focus to feeling the quality of energy radiating from our heart and spreading throughout our being. It is life-enhancing unconditional love, connecting us to all beings in the universal consciousness of the Creator.

By aligning ourselves with positive, rising vibrations of joy, we can open ourselves to expanding awareness within ourselves and all creatures that we focus upon. These are subtle vibrations, and in order to be aware of them, we must be present and clear of attachments to limitations. We can invite them into our awareness and accept the vibrations that we receive, as our consciousness interprets them into communication that we can understand.

Because we have agreed to participate in human life here, we compartmentalized our awareness at birth to the consciousness within time and space in empirical reality. For the real human experience, we could not allow ourselves to know our true, unlimited Self, or we would know that we're just playing at being human. We couldn't have the deeper experience of living in duality. Once we begin to realize that we are much greater than our human person, we can transform our lives into experiences of love and beauty.

Our realization comes from our inner knowing, which our ego-self largely discounts. Our intuition is not a mental process. It is a feeling of alignment with an understanding and a deep knowing. It is accompanied by our inner sound current, which changes with our level of conscious awareness. This is all mysterious to our ego-consciousness, which can recognize only limitations. As long as we are attached to our limitations, we cannot truly follow our intuitive knowing or realize our greater Self.

The entire world that we have considered real in our experiences is a limited portion of our consciousness. We are much greater and can awaken to our infinite Being by resolving our limited beliefs about ourselves and opening to our intuitive knowing, which is present for us in every moment through the energy of the heart of our Being.

Moving from Ego-Consciousness to Intuitive Guidance

As we move through the turning of the ages, the hidden, most negative energies that have existed among humanity for eons are being brought into the light. As a result, our society appears to be crumbling and destroying itself from within. This is a healing catharsis, moving us to understand that we are much greater than we believed. If this were not happening, the negatively-oriented beings could have completed their enslavement of humanity with no hope of our ever realizing the truth of our Being, because people would not have realized that the prison doors were closing permanently.

Although the mass of humanity has acquiesced to living in duality, recognizing suffering and mortality as inevitable, it is only a choice, and its reality in our experience is created in our own consciousness. We are not naturally required to live in need of anything outside of our own being. We all are designed to be masters of our human lives, living in joy and confidence, free to be whoever our hearts desire.

Now the light is dawning within the consciousness of many. When we become only positive, guided by our deepest inner knowing of positive, high vibrations, we can live in confidence and freedom that cannot be believable for those who are unaware of their intuitive guidance. There is more than one level of life present among humanity, and we have the ability to choose the one we want to live in. We can live in duality or only positive. These are experiences of conscious recognition and realization.

As we become more capable of expanding our awareness in greater vitality and joy, our lives become more abundant and richer in every kind of experience. As we resolve our limiting beliefs and are able to transcend ego-consciousness, we can more fully realize our inner knowing and divine purpose in relation to all conscious beings. We become capable of living beyond the negative energies present in humanity. We can be aware of conflict, chaos and life-diminishing energies, but they exist in their own realm of polarity and vibratory level. We are not required to interact and participate in that spectrum of energy.

We can actually live in a level of energy in which we do not need to encounter negative energy, once we become consciously and subconsciously completely positive in alignment with our higher-conscious intuition. At this level of consciousness, we are also in alignment and telepathic communication with ascended beings beyond the physical. Our intuition protects us from being in the wrong place at the wrong time. The transition from ego-conscious guidance to intuitive knowing is the challenging part. Once we know and practice the process of intuitive alignment, we can make the leap in consciousness more easily.

Aligning with Gaia's Rising Resonant Frequencies

Because of the rising resonant frequency of our planetary home, the level of our personal vibrations must also rise. If we are out of resonance with Gaia, we will be very uncomfortable and probably won't want to be here. We are moving toward more positive personal experiences, and the parasites are having to leave. Our near future includes a transformation of everyone who is open to greater love, and the others will be uneasy and stressed-out.

The geomagnetics of the Earth are changing, and scientists are predicting that our planet is losing its negative pole and shifting to a single positive in the center. This supports the rising resonance and indicates that we are moving toward only positive vibrations in a higher spectrum of energy. It is the realm of love, compassion, gratitude and freedom.

Our most important form of freedom is our freedom to choose our perspective, our thoughts and our emotions. Having control over our thoughts and feelings and being able to realize their presence when desired, enables us to be trustworthy to ourselves. When we choose to connect strongly with our intuition, we can transcend our personal limiting beliefs about ourselves, because we can experience expansion of our awareness beyond the limitations of living in duality.

If we are interested in being only positive, there are things we can do to take us in that direction. We can practice deep, slow, rhythmic breathing in order to relax into a more expansive state of being, as we relax our focus and become our unencumbered present awareness. In this state of being, we can align ourselves with gratitude and joy.

In our encounters with others, we can accept everyone who comes into our presence with gratitude and compassion. There may be some rocky encounters for a while. These we can use to learn greater control of our mental and emotional alignment. The more we practice our own form of deep meditation, the more we can realize our infinite Self, beyond our accustomed human experience. When we spend time alone and barefoot in nature, we can sense the rising resonance of Gaia and align with it. She is a Being of unconditional love for all of us, regardless of what we do to her.

For us to be able to experience unconditional love, we must align with our deepest inner knowing. It takes us into the heart of our Being, where, in gratitude and joy, we can sense the presence of conscious life force arising within universal consciousness and modulating the energies that pass through our mental and emotional states of being into alignment with us. By being sensitive to these positive, high-vibratory energies through our intuition, we can continue to expand our awareness.

Conscious Transcendence

Designed to provide experiences in duality, the world of human experience contains an undercurrent of fear of suffering and termination, because we believe in the death of the body. Negative beliefs have led us to subject ourselves to the dictates of parasites, because we've allowed ourselves to be intimidated. We have been trained to believe that we must live within the structures of our society, when, in our true Self, we are infinitely creative.

Our ego consciousness does not know this and cannot believe that we can create everything we need and desire. Because we are by nature energy modulators, our beliefs filter all of the energy that we are aware of. When we believe that something is real, we can experience it as real.

The designs of our traditional experience are changing, and material cause and effect is not necessarily how things happen now. We have learned that our consciousness is the origin and creator of our life experiences. We are either positive or negative. Humans generally are negative to some degree, because we have this undercurrent of fear. Until we decide to transform, we are restricted to the matrix experience that we have accepted as real. This is the world of human experience, limited by our beliefs in our mortality, the necessity to work for money, and a feeling of lack of love and joy.

Unless we choose it, we are not required to subject ourselves to any of this. We have the choice to expand our conscious awareness. This happens by aligning with higher vibrations that are positive. We can be aware of these in the energy of our heart. Our heart works tirelessly to enhance our vitality to the extent that we allow, by how we think and feel about everything in every moment, and how we care for ourselves. Our heart cannot be intimidated.

By asking within ourselves for higher guidance and then being open to receiving our inner knowing, we can learn to align with our intuition. Our real breakthrough happens when we can realize our eternal present awareness in great love, gratitude and joy, which are the expressions of the consciousness of our expanding Self, drawing to us experiences that stimulate those feelings. This level of consciousness transforms our lives and enables us to be Self-sufficient in every way.

Reflections of Higher Consciousness

We can be poised on the limits of human awareness, wondering about the potential of our consciousness and how to enhance our awareness of it. We can feel and know if our personal vibration is positive or negative in any moment. We can sharpen our awareness and penetrate the energy patterns that we confront with our own energy signature. We feel and know when something resonates with us. This is attractive energy.

As we move through our life experiences, every encounter has potential outcomes that change with our own emotional polarity. We are the directors of the quality of our own energy in every moment. Each vibratory pattern that we align with in every moment reflects its own energetic level and creates qualities in approaching energetic patterns of empirical experiences.

There is a natural flow of life that we can align with, and we can know the quality of vibrations that we choose to focus on. We can intentionally choose to be aware of positive, high vibrations in every moment that is possible for us. By keeping our vibrations high, we enhance our life experiences in every moment in which we are positive. Any departure from resonance with our energetic essence results in instability and limitation of our consciousness. We tend to block our awareness from what we do not want.

On an energetic level, we tend to block our awareness of polarity and vibratory levels beyond our personal resonance. Our experiences are filtered through our acquired perspective. All of our limiting beliefs are based on some level of fear. If we can recognize this and accept it with deep understanding and compassion we can resolve our personal attachments to our beliefs in limitation, and we can be clear in our understanding of each moment.

Once we are mentally and emotionally clear, we can realize our intuitive guidance in so many ways. It can come to us in visions, feelings that arise, writing, hearing an inner voice and even what we speak, signs, photos, and many more ways that subtly direct our attention to more loving and joyful vibrations that we can align with and express, elevating everything around us.

When we are awake to these inner promptings of guidance and enhancement, we naturally become more compassionate, grateful and free. Sensitivity to our intuition is all we need for masterful guidance in joyful living.

Guidance to Greater Awareness

Aligning with the energetic quality of our intuition opens us to our higher guidance and transforms our lives. Achieving this alignment happens when we focus our attention on positive, high-vibratory patterns of energy. We can imagine being in experiences of beauty, majesty and joy. Then we can let our intuition provide the visions and feelings, while we remain clear of personal interests. It may only take a moment of realization to know something important.

In order to receive guidance from universal consciousness, we can align ourselves with positive, high-vibratory compassion, love and gratitude. This life-enhancing energy is the vibratory quality of our intuitive guidance. It directs us to realize the truth about who we are as beings of unconditional love and joy with infinite creative power. Our human being is only a limited expression of our true Self. Through our intuition, we can open our awareness to our greater consciousness.

Our conscious awareness is not temporary, even though we have set boundaries for it to give us a genuine experience of duality, with the ability to feel negative energy. This is the temporary part. In the limited spectrum of the empirical world, we can experience destruction, but the end is only the end in the empirical world, not in our expanded reality of eternal awareness.

We are naturally only positive, because we know our essential being is an expression of unconditional love in the Creator consciousness of the Being who is expressing Itself as us in our current form. We are the unlimited Creator in all of our forms, including our limited, human expression. Within the empirical world, we can experience ourselves as separate beings, apart from the consciousness of the Creator.

Only in this way could we ever stare into the depths of darkness the way we have and survive, providing ourselves with greater wisdom and compassion. The Earth-human experience is a great challenge, generating incredible fear and expected termination. Through intuitive guidance, we can remember our true Self. In learning to align ourselves with our intuition, we begin the journey to higher consciousness in a way that transcends our limitations and allows us to resolve them.

Once we realize that we are unlimited, we can release our limiting beliefs. While we are still stuck with our limitations, we can choose to align with our intuition and be guided to greater awareness.

Encountering Our True Selves

In realizing the true essence of who we are, we have many opportunities to play with our consciousness. Human life is a game of consciousness restriction and expansion, filled with wild emotional and mental challenges that we take very seriously, believing them to be real. We have expertly developed our ego-consciousness to operate within the confines of our beliefs. So convincing have we made our empirical experience that we don't even imagine anything more as a possible reality.

Only by being curious about life beyond our beliefs can we expand beyond our human reality. If we look for clues about how to do this, we can become aware of the nature and essence of our personal human presence. It is our present awareness that enables us to command our ego-consciousness from a place of deeper knowing. By being aware of the inner guidance that we constantly receive, whenever we focus on it and align with its energetics, we can identify the level of positive vibrations we can follow to higher conscious awareness.

We have learned to participate in a very complex and convincing game of restricted consciousness, designed to enrapture and capture us within its spectrum of duality and empiricism, manifested for us through our beliefs in our limitations and our donation of conscious life force through our fear of suffering and termination. Without our life force and energetic alignment, the human matrix could not exist. We are its creators and rightful masters.

Through our realization of the nature of human life, we can open ourselves to physical and imaginary experiences of greater awareness and understanding. We can begin to know beyond belief what is real. We can offer our ego-consciousness the scientific understanding that what we experience in our bodies is a very limited spectrum or bandwidth of electromagnetic waves and patterns of waves that we have adapted to perceiving with our physical senses.

There are realms just as real in infinite consciousness, available for our recognition when we can resonate with their energetics. We have an emotional intelligence that is aware of any vibrations that we focus on, as well as a large spectrum all around us. We receive as much emotional stimulation as we believe is real. As we begin to open to greater awareness, we receive visions that we believe are imaginary, until we start to feel their energy. Once we can feel their presence, they become real for us, and we intuitively know who and what they are. In this process of developing inner sensitivity, we can follow the feelings that we request into greater awareness of our expanding Self and our true creative ability.

Awakening from Our Trance within Space and Time

On the path to full awakening from the hypnotic trance of humanity, our goal is to resonate with the universal consciousness of the Creator of all. We can recognize that we are fractals of the Creator, endowed with infinite creative ability. Access to the abilities that we have blocked ourselves from comes through our Intuition in the energy of our heart. Our intuition is our conjunction with universal consciousness. Intuition is its own proof of what we know without any other proof. It is always life-enhancing and compassionate, and it guides us in the way that we choose. When we search for it, it is always present in a positive way for us to be aware of.

In order truly to open ourselves to our potential, we can transcend our limiting beliefs about ourselves, by intentionally focusing on scenarios filled with joy and love, while choosing not to pay attention to thoughts and feelings that diminish life. We can set our preferences to the greatest brightness, and be able to feel it, even in the darkest of beings. They also are us in the deepest aspect of being, where we have shut a supremely fearful experience into a private compartment of our consciousness. Unavailable to our conscious awareness, the unresolved, deep-seated feelings of disconnection from life still haunt us vibrationally and keep us from knowing our true essence.

By realizing that we are free in our awareness, we can understand how limitations diminish our lives. They are manifestations of fear and are the opposite polarity of love. True love is unlimited, as is our awareness. When we are in resonance with universal consciousness, we can realize that we are our eternal, presence of Self-awareness, apart from our physical identity. Our physical body is a manifestation of each moment of our energetic presence, the alignment of our thoughts and emotions.

Since the universe is always balanced, every action that we take is what we are doing to ourselves. This is what is known as karma. We experience sending the energy out and receiving it back into our experience. Being always positive in the energy we send out, brings compatible energy back to us. As we are able to maintain a positive perspective of brightness, we gain the ability to transcend our limiting beliefs. Without beliefs, we can be guided by our intuitive thoughts and feelings. This attention enables us to be lavished in creative manifestations of the natural desires of our heart.

When we can realize that we are the creators of the energies that form our experiences, we can learn how to align with the energies that we want to experience, including energies that we have considered impossible for us personally, like higher mathematics, telepathy and thought travel. Since all of our limitations are self-imposed, we can transcend them by paying attention to the qualities, in all of their richness and subtleties, of our intuitive guidance in each moment.

Transforming our Enslavement into Unlimited Freedom

By releasing our attachment to energies that we don't want to experience, we can stop giving them reality. For them to be real for us, we must recognize them and believe that they are real. If we do not give them our attention and belief, they disappear from our experience. Releasing our belief in their reality requires deep introspection in search of the source of our infinite Self-awareness. This is the essence of our consciousness and is who we are in our immortal Self.

Beliefs in our personal limitations are self-constructed through parental training and social conditioning, and we can deconstruct them through positively intending to express our true Self. Ego-consciousness begins to disappear into the background. In its place arises our intuitive knowing. As our inner feeling and knowing become a constant presence in our awareness, we can release our fears and feelings of insecurity and stress. These are accompaniments of ego-consciousness and depend for their reality upon our limited awareness.

If we search within our own awareness, we feel anchored to our physical presence. We can sense many kinds of feelings as our consciousness envelops all parts of our body, but our awareness is not limited to the empirical world. We also have our inner feeling and knowing, our intuition, which we can become sharply sensitive to.

As we learn to be intuitive, we begin to transform our lives and to align with our greater identity. Once we learn to depend upon our intuition, we can face our challenges with confidence in our connection with infinite knowing. At any time, we have the choice of opening our awareness to the wonder and majesty of who we really are as fractals within the consciousness of the Creator of universes. We participate in the consciousness of the Creator as extensions of the Infinite One and Creators of experiences that express the qualities of our state of being, which is a result of the energetic patterns of our thoughts and emotions.

The true function of our free will is to play intentionally with the polarity and vibratory patterns of the unified quantum field of all potentialities. By our perspective and intent, we change the qualities of the electromagnetic wave patterns around us. We are designed to express the intent of the Creator through our own internal mental and emotional processes. By intentionally holding the unconditional love that flows through the heart of our Being, we can express the enhancement of all of life.

Our Experience of Consciousness. Thought, Belief and Realization

Consciousness is universal. It is the essence of the Creator of all and is the creative basis of everything that exists. It is present in everything and envelops everything. It is the source of our life, and it provides us with our personal identity. As quantum physicists have shown us, this is true for all conscious beings, even the smallest sub-atomic particles/waves. We participate in universal consciousness as much as we allow ourselves to know this.

Just as any part of a hologram is a fractal of the whole and contains the whole, our individualized consciousness is a fractal of the whole. Our consciousness enables us to be infinitely creative through our use of the creative aspects of our being. We have a direct connection with universal consciousness through the intuition of our heart. Intuition is not a mental or emotional process. It is direct realization, a kind of knowing that has an energetic quality that feels like life-enhancement and unconditional love. It is elevating and inspiring and can be unerringly available to us only when we are open and attuned with it. If we want guidance for the true path of our life, this is where we can find it.

Because we are individualized Creator consciousness, we have free will to restrict our awareness by creating limiting beliefs about ourselves. We have created our ego-consciousness, which operates through our beliefs to enable us to have the full experience of living in duality in apparent separation from our full consciousness. It is a very convincing experience, so convincing that we have become entranced with the power of negativity in our experience. Duality and negativity are our own creation. Without our recognition and belief in the reality of the negative, it could not exist. It has no conscious life force from the Creator, only from us through our engagement and alignment with its polarity and vibration. It is created and held in the consciousness of humanity, and we are the ones who can dissolve it.

If we release our own belief in our limitations and in the reality of two powers of opposite polarity, we can free ourselves to realize our true Being. Our limiting beliefs are deeply set in our awareness and are resolved with great difficulty, but if we are strongly intent on realizing our truth, we can succeed in becoming the masters of our lives by opening ourselves to the wonders and richness of our intuition.

Understanding Unconditional Love

The ancient Greeks recognized three different kinds of love. One was erotic love based on sexual attraction and the desire to create more lives. It is part of our ego-consciousness. Another was the love between friends and family based on mutual interests and genetic connections. This is also part of our ego-consciousness. The third was called *agape*, often known as spiritual love, which transcends personal interests and is based on our divine essence. This is unconditional love. It is not an emotion or feeling; rather, it is a state of being or a level of conscious vibration that enhances all of life. On an emotional level, it is related to joy, compassion and gratitude.

Unconditional love is a conscious sharing of our life force. It is based on deeply knowing our unlimited Source of life in our eternal presence of Self-awareness. It is knowing that we are all the same Being, expressing Itself as each of us. All conscious beings exist in the universal consciousness of the Creator of everyone and everything. In our essence, we are the Creator. We are fractals, fully endowed to create universes. This is all part of unconditional love, which expresses the essence of the Creator.

Because we have free choice, we can create anything that we desire to experience. When we desire to be in a state of unconditional love, we can open ourselves to knowing and feeling the presence of the Conscious Creator enveloping us and feeding us an infinite stream of creative life force for us to create more experiences that inspire and uplift all conscious beings, while enhancing our own lives as well.

Even evil people are expressions of unconditional love in their essential being. To their own detriment, however, they have used their free choice to align with and create life-diminishing experiences. As a result, we can have deeper compassion for all who experience negative polarity. We can recognize the divine origin of the dark ones and be compassionate for their energetic sacrifice, cutting themselves off from their life force and their experience of unconditional love. To make up for this, they steal life force from others through psychopathic relationships, but they cannot steal unconditional love.

As divine Beings, we feel most comfortable in the state of unconditional love. Our intuition is our guidance that keeps us at this level of energetics. Intuition is the communication of unconditional love through our own aspects of awareness. Opening ourselves to our intuitive knowing is an intentional act and is how we can transcend our limiting beliefs about ourselves.

Aligning with Universal Consciousness

We can expand our awareness into the consciousness of the Creator, whose presence is everywhere and in everyone. Most of us don't seem to recognize it, and some have almost completely blocked it out of their lives. As a result, they have become life-force parasites. The presence of Creator consciousness is what gives us life and unconditional love in eternal awareness. This is beyond ego-consciousness, which is limited by nature and is no longer useful in eternal awareness.

To be aware of being in the consciousness of the Creator, we must raise our vibrations to the level of gratitude and just be aware. It's best to be in a serene natural environment. We can invite the energy of our heart to come into our awareness. This is our connection with universal consciousness, the consciousness of the Creator. It is also our intuition, our inner knowing.

As we realize that our awareness is expanding, we begin to recognize the presence of other awarenesses. We all share the same consciousness, and it is everywhere. All conscious beings arise out of the same consciousness with our own individual awareness. Even though we have chosen to express ourselves as separate bodies, we are all the same Being, connected in unconditional love and vitality, and living in multiple dimensions, separated by energetic polarity and vibratory patterns.

By opening ourselves to positive energy, we align with universal consciousness, and we're able to focus on elevating energies. The more positive we become, the better our lives express love and compassion. We become helpful and life-enhancing to all around us. All things that our hearts desire come to us without effort on our part, unless we're intuitively guided to do something. We become Self-motivated without ego-consciousness.

Intuition is higher guidance, but it is not our master, the way the ego has attempted to be. It is not intrusive. We have our free will and can decide in every moment how we want to feel and think and what energetics we want to align with, but we do need guidance to play our roles easily and with confidence that we are expressing our destiny in alignment with the consciousness of the Creator.

The Importance of Just Being Present in Awareness

The continuing task of all who are on the inward path is to learn and practice being our eternal, presence of awareness, just being open and aware, allowing the energies of Gaia and our enveloping cosmic field to reveal themselves to us. In pure awareness we have no attraction or resistance to anything. We just are present and aware, letting the energies that we encounter pass through us without judgment or alignment. They are just patterns of energy, capable of stimulating feelings and thoughts in us, attracted to us by the quality of our energy signatures. These have been mostly creations of our ego-consciousness, within the bounds of the energetic patterns held in the consciousness of humanity. The entire world of human experience is made real for us by human life force through our recognition and belief in its reality, and our resonance with its spectrum of vibrations.

Although it's just patterns of electromagnetic waves organized by polarity and frequency levels, the world of human experience in empirical duality appears very solid. The entire world that we perceive happens only in our own consciousness, and it keeps entering our thoughts and emotions, even when we just want to be clear and present in awareness. These energies that enter our awareness do not need our alignment, and we can let them pass through, as if they were scenery on a voyage through time.

The awareness that we are being attracted to is always present within our heart as the inner guidance of our intuition. It offers no judgment, only support for our enlightenment. It is our connection with universal consciousness and comes from the Source of unconditional love. It confers joy, freedom and abundance throughout our life, when we can appreciate and receive its guidance through our alignment with its polarity and vibratory level.

In order to benefit from our intuitive knowing, we just need to be present and aware as much as possible, as well as mentally and emotionally clear. We can feel the warmth of energies that we are naturally attracted to. These are empowered by divine intent for the creation and enhancement of all life. In order to align with these energies, we first need to release our attachments to limitations.

All of this requires practice in resolving our subconscious training and fixations, which arise in our awareness, when we begin to move beyond their limits. This is when we can examine them and realize if they have a basis in love. If we direct our awareness to the energy of our heart, we receive abundant love and joy. This is our creative power, and it is unlimited. All we need to do is be aware of it and realize what it is.

Being Aware of Our Radiant Presence

We may live in a higher dimension of consciousness, but we still also participate in the world of duality as held in the consciousness of humanity. We are creating a world that is only positive, and we participate in it whenever we can imagine it, feel it and realize its reality. In our inner knowing, our intuition, we can pay attention to only positive and true promptings. These are part of our awareness, like our conscience. We receive often subtle senses, prompting us to do something or be aware of something. We can learn to trust these feelings and directions and let them guide us in navigating our lives. This is a step beyond ego-consciousness, requiring resolution of our self-limitations, so that we can be clear in our awareness.

As our awareness expands, we realize our multidimensionality. We can be aware of the radiant presence of everyone, both embodied and purely energetic. We can choose the level of light and emotional vibration that we prefer, as well as the kind of polarity. In making these choices, we align ourselves with others who resonate with us. In our own awareness, if we do not pay attention to negative energies, they disappear from our presence, except for those that we decide to transform in our imagination and feelings coming from our heart.

In our personal encounters, we may intend and be guided to enhance the lives of everyone, including ourselves. Everyone can be enfolded in our love and compassion, which come from an infinite Source, enhancing our lives as it flows through our consciousness.

We may encounter negative energies that attempt to force us into compliance with their vibrations, but they have no ability to do that, if we maintain a positive perspective of enhancing the lives of everyone. If the directed energy is negative for anyone, it is based in some level of fear or greed. Without our engagement and permission, it cannot affect us, once we realize our eternal presence of awareness beyond the body. No energy can affect us, unless we align with it. Negative experiences happen to us because we fear them. If we do not hold fear in our attention, its results do not manifest in our lives.

By aligning our intentions with the energies of the heart of our Being, we expand our awareness beyond our self-imposed limitations in time and space, as well as duality. We can transform our human presence into realizing and expressing our infinite Self.

Transforming Our Personal Energy Signature

In order to be the masters of our lives, we can learn to disengage from negative energies in our own psyche and transform our vibrations into love and compassion. Because we share in the consciousness that creates and envelops universes, our primary concern is our own consciousness. If we dislike someone or some circumstance, we can transform the energy that we perceive by transforming ourselves.

We are channels of the conscious life force arising from the Source of our life. We are endowed with awareness able to control and direct our thoughts and emotions. Through practice, we can achieve this. We are also endowed with a conscious connection to the energy of our Creator. Once we have resolved our attachments to limited, negative beliefs about ourselves, we can align with any level of energy that we choose. We can recognize divine energy and choose to align ourselves with it. Our negative beliefs have kept us from realizing the positive vibratory levels of our Source energy, which is at the level of joyful ecstasy in Self-Realization and creative power. In this state we are able to play freely in any way that aligns with the energy of our heart. By maintaining this focus, we are able to realize ourselves as pure conscious presence of expansive awareness, having our own state of Being and energy signature.

We can pay attention to anyone and anything we choose. By choosing our polarity and vibratory level, we determine the quality of our personal experiences. Although we live in a world that we share with all of humanity, and with lots of negative energy being expressed, we do not have to experience any of that, even though it may be all around us. It does not have to be within us, but because of the distractions all around us, we must practice mental and emotional control and direction in every moment. This requires strong intention, but we have free will. Consciously or subconsciously, we choose the focus of our attention, always manifesting in our state of being.

We can be the masters of our inner knowing and our creative envisioning, guided by our intuition, our connection with our Source-consciousness. This energy is what we radiate all around us. We attract experiences that resonate with our state of being. Although it may appear to be, this has nothing to do with our larger environment. In our consciousness we can disengage from the realm of duality and transform into a higher-vibratory dimension, unaffected by the energetic quality of negativity.

Because of our creative essence, we personally experience the energetic level that we align with, and that we share with the consciousness of humanity. Our tools of transformation are our thoughts, words, feelings and how we live and express ourselves. With our intuitive knowing as our inner guide, we enable ourselves to channel the awareness of Source energy. We can live in deep gratitude and joy. As we focus on maintaining a level of feeling love and compassion, negative energy disappears from our lives. It no longer receives our life force, and as it weakens, we become stronger, unaffected by negative energy in any encounters.

The Secret We All Hold Within

Created in unconditional love in the consciousness of the Creator of all, we are given our own unique eternal, personal essence, with the freedom to express ourselves however we choose in every moment and in any dimension and density. In our essence we have unlimited awareness and creative ability. As fractals of the One consciousness, we share our awareness in deepest love and joy with all the elements, all conscious beings, animate and inanimate, and all the entities of nature.

To the extent that we have chosen to experience life as humans on this planet at this time, we have entered a kind of hypnotic trance and imposed upon ourselves many limitations, including participation in dark energy in an experiment to know what it is like to separate ourselves from awareness of our Source. We have chosen to operate within ego-consciousness and without higher guidance or knowing our true essence. We have delved so deeply into negative energies, that we are now in danger of destroying ourselves and our planet.

It is time for humanity to awaken to the essence of who we really are and return to our full consciousness as immortal Beings of great light and unconditional love. This capability lives within each of us in the presence of our intuitive knowing in the heart of our Being. This is our higher guidance. We have intentionally not been paying attention to it, and it has become dim, but it is always present. To become aware of it, we can learn to control our breath, our mental processes and our emotions. We can learn to become serene within and to listen and be aware of what we deeply know in every moment.

We can call our guides and angels and other ascended Beings to draw our awareness to our true essence and to inspire us to expand in unconditional love, gratitude, compassion and joy without limit. Every situation and every person displays the life force of the Creator, or they could not exist. If we look for the light in every encounter and intend to interact only with it, we can recognize the presence of the divine.

The more positive we can become, and the higher we can go in gratitude and joy, the stronger our intuitive awareness becomes, ultimately lifting us into a realm of ecstasy, freedom and beauty unimaginable to our ego-consciousness. As our connection with our heart-consciousness strengthens, we transform our lives into experiences of miraculously wonderful encounters and feelings, regardless of what seems to be happening around us. By paying attention and aligning our awareness more completely with our intuitive guidance, we can release our attachment to limitation and can transcend our ego-consciousness. Awareness within universal consciousness awaits our realization.

Learning to Live in the Consciousness of the Creator

If we allow ourselves to open our awareness to the greatest love and joy of our expanded Self, in alignment with the energy flowing through our heart, we can clear our attention beyond our limiting beliefs. We can imagine what those feelings are, and we can practice summoning them. The more we do this, the clearer we become and the deeper we can go in love, joy, compassion and gratitude. These are very high vibrations. By focusing on scenarios filled with them, we can transform our perspective and our life experiences.

By being free of personal needs and addictions, we can expand our awareness beyond our current simulation of reality. Our penetration of consciousness can expand our awareness and enrich our lives without end. Usually this kind of awareness comes after much practice in deep meditation. Once we experience it, we are no longer bound by anything. We are just clear awareness, knowing our own presence and able to vibrate in unconditional love in every moment. We can be aware of every other presence around us and know if they are in resonance with us. We can also be aware of what is about to happen around us, especially significant things in our near future.

This level of living is our destiny. It's how we are created to be. To get there from where we have been requires effort on our part. Control of the mind and emotions in alignment with the energy of the heart of our Being allows us to trust ourselves to create experiences that we truly want. Vibrations on this level enhance the lives of everyone and radiate into the quantum field and through all conscious beings. Our conscious presence extends far beyond our human experience, and is available for our realization in every moment. It comes to us through our own inner knowing.

Each of us can transform our lives to our highest and best potential by doing our best and intending to continue to expand in awareness of unconditional love. This is our natural way of being and is how the cosmic energy through the Sun and the Spirit of the Earth are drawing us into resonance with it. Many beings without a physical presence are also influencing us to realize a greater life and more expanded way of being. We can be aware of all of this through our intuition.

On an energetic level, by far the most powerful organ in our body is our heart. It has greater mental ability than our brain and is much more powerful in its radiance and influence around us. It speaks to us through our intuition and offers guidance from the consciousness of the Creator, which we share as conscious Beings. In our ego-consciousness, we cannot realize this, and we remain skeptical and doubtful. When a strong interest arises in us to know what our life is about, it's time to recognize our intuitive knowing, which can happen by intentionally raising the vibrations of our state of being. We do this by focusing on elevating our awareness and learning to see the light of the Creator in everyone and in every encounter. With our energy signature, we modulate the energy in the etheric realm, eventually manifesting into the quality of our physical experience.

Releasing Our Parasitic Controllers

As our psychopathic controllers work to make things as unbearable as possible for us before they cannot bear to be here any longer, humanity is suffering in the present moment. For this we can have deep compassion and be moved to create experiences of abundance and freedom. The outer world does not have to affect our own vibratory level. By staying at the level of gratitude and compassion, we can envision and emotionally participate in imagining the experiences we want to create. Creation results from our energetic alignment. When we are aligned with the intuition of our heart, we are creating life-enhancing experiences.

Prompted by our intuition, we can adjust our imaginary energetic alignment to positive, high vibrations of gratitude, love, compassion and joy. By maintaining this perspective in every encounter, we help to destabilize negativity and enhance positivity, which is the inherent energetic structure of everything in the higher dimensions. If we do not engage with negativity in resonance or resistance, we do not create it. It receives no life force from us, weakening its existence. It is artificial, created by humanity. It is just a distraction from the creative focus of attention that our inner journey to mastery requires.

This means that we need quiet time to meditate and listen within. We need to be grounded to the Earth and receptive to the Sun, the brightest and most radiantly powerful Being in our experience. In natural environments, we can align with the feelings of all the plants, animals and rocks. They are all conscious beings expressing themselves as they are. Their vibrations are naturally in alignment with the energy of the Spirit of the Earth. It is from a place of serenity that we can most easily expand our awareness into a higher vibratory dimension of living, and we can learn to maintain this level of present attention.

For a while, we may encounter great challenges to a perspective of gratitude and compassion. We can remember that we cannot be harmed in any way without our consent at some level. This kind of consent depends upon our having a victim attitude. Again, it's a matter of energetics. If we engage negativity in any way, we are creating it. By staying in high vibrations, we do not align with negativity or allow it to have access to our life force. When we achieve a perspective of only positive, we no longer create negative encounters.

Chapter 2. Guidance on the Inner Path

Choosing the Quality of Our Life Path

Whatever is happening contains the clues to awareness of an elevating energy that is drawing us toward more expansiveness and greater love. If we choose to focus our attention on that level of energy, we can be mentally and emotionally attracted enough to vibrate at that level. We can choose to be either positive or negative. Neither one is better or worse than the other. They just provide different qualities of experience. We've been down both paths, and we know the kind of experiences that they offer or impose. Now, in complete freedom, we can choose for ourselves the spectrum of vibrations we want to align with, regardless of what may be happening around us. We know the kind of experiences we want. If we decide to go the positive route, our experiences will be very different from the negative route.

The negative route has no intuitive guidance. Higher guidance works when we choose to be only positive, anchored in unconditional love and compassion. Here there is an abundance of joy, and there's an easy flow in living. Conditions are supportive of everything that enhances us and all of humanity and beyond.

It is the nature of the path that we're on to provide experiences that resonate with its energetics. We co-create with the energies that we focus on. The qualities that we are aware of come to us filtered through our limiting beliefs about ourselves. If we critically examine our beliefs, we can come to their origin and accept them with compassion and love, resolving them.

Positive energies do not have limitations, as opposed to negative energies, which are limited by potential dissolution. Positive energies are regenerative and enhance the vitality of everyone. They result in fun and enjoyment. The level of our ecstasy is infinite. In this state of Being, we can know unconditional love, gratitude, joy, freedom and abundance in every way. For every moment that we can be completely positive, our vibratory spectrum radiates our energetic pattern into the quantum field, out of which arise the qualities of our life experiences in the forms appropriate to our environment.

Here's a vision that we can manifest for anyone who loves having a body. We can learn to vibrate at a very high, positive frequency and train our physical cells to resonate with us, raising the vibrations of our physical bodies, so that we can still have the enjoyment of physicality in a higher dimension, where everything is positive. We can realize our physicality as essentially a vibratory pattern that resonates with infinite Being.

Learning to Live in the Light of Our Heart

Whatever we think and feel has an electromagnetic vibration that shapes our personal energy signature, which radiates into the quantum field that envelops us, attracting energetic patterns that resonate with our own. Although we are accustomed to having thoughts and feelings flow through us according to patterns in our ego-consciousness, we can intentionally direct our thoughts and emotions through the focus of our attention. What is important is the polarity and vibratory frequency of our focus.

If we are negative, harboring anger or fear, we can imagine experiencing things that we want, but they will come into our lives along with unfulfilling challenges to our well-being. Positivity creates positive experiences. Everything begins with our own perspective and state of being. We are constantly being creative. It is our nature and the essence of who we are. What keeps us from realizing our creative ability is our attraction to doubt.

In order to manifest the fulfillment of the desires of our heart, we must be in alignment with its vibratory energy. This is where we realize what we truly know. Since the process of knowing is different for each of us, we must develop our own intuitive sensitivity by focusing on inner awareness. This involves making peace with our ego-consciousness, so that we can relax into a state of serenity, listen to our inner sound current and ideally find the space between waking and sleeping, when the ego is silent. This is where we can expand into our infinite Self.

Much intentional practice may be required to achieve this level of Self-Realization, but along the way we can realize great personal transformation, guided by the energy of the heart of our Being, which comes into our awareness when we seek it, ask for it and are open and receptive. We can direct our awareness into alignment with the most positive, wonderful experiences that we can imagine, and then hold this focus, as we allow unconditional love and gratitude to flow over, around and through us.

When we can remain positive in every circumstance, we experience miracles and great vitality. We open the way for a life filled with abundance and joy. This is our natural state of being. It is a state of complete confidence in love and fulfillment for ourselves, as well as all conscious beings. It is the radiant and magnetic energy of our ascension.

Transforming Ourselves with Our Bodies

To progress toward Self-fulfillment, we may want to learn to interact lovingly with our subconscious self. It is the part of us that is our faithful servant in providing everything we command with our vibrations to arise in our bodies. It is our body consciousness, directing every life process in all of our cells, so that our bodies resonate harmoniously with the vibrations of our state of being. If we are only positive, our subconscious will adapt and remove the negative defects in our bodies.

Our physical defects are symbolic messages to us, asking us to become aware of the negative energies that we have adopted. As we examine any parts of us where we feel any amount of fear or doubt, we need to be objectively honest with ourselves and accept what we feel. Do these energies benefit us in any way, apart from fear of a negative outcome? Anything that does not resonate with our heart-energy is a defect in our auric field. With continuing personal monitoring, we can discover our spiritual lessons and choose to transform their causes by bringing our vibrations into alignment with our heart's intuition. This is where we can understand and know our situation and receive higher guidance. It is all within our own consciousness.

Being pure of heart, and having deductive reasoning ability, our subconscious does not understand us, because of our inductive abilities, but it can read our energetic polarity and vibratory level, and this is what manifests in our bodies. There are many things we can do to correct the causes of our problems. One of the most powerful is past-life regressions. Taken all together, our personal defects and limiting beliefs about ourselves are all under examination. If we are aware of the quality of these energies, we can transform them into positive, life-enhancing scenarios, allowing our subconscious self to align with our heart in every moment and to perfect our bodies.

By caring to communicate with our deeper self, we can express our gratitude and love for our perfect body, while imagining ourselves being how we desire. By practicing this every day, and being aware of our energetic state of being, we can come into alignment with our higher knowing. Our entire being can come into alignment with universal consciousness through our intuition.

The Magnificence of Who We Are

There comes a time for all of us when we suspect that we may be more than mere humans living in subjection to the energetics of empirical duality. We begin to question the boundaries of our consciousness. If we can look at them objectively, and even logically, we find that they are all based in some kind of fear. They are negatively polarized. Once we realize this, we can understand that our boundaries are artificially contrived in order to provide the kind of empirical experiences that humans can have.

In our true essence, we are fractals arising within the consciousness of the Creator, gathering experience and understanding for infinitely-expanding universal consciousness. Our every thought and feeling have unique vibrations, and are part of universal consciousness. We are constantly created and held in the eternal consciousness of the Creator. We are given power to create experiences through our mental and emotional actions and reactions, and we are given inner guidance in knowing beyond limitations. Fear does not exist in our inner guidance, which is rooted in unconditional love and radiant life force.

By opening ourselves and being receptive to our inner knowing in every moment, we can practice being sensitive to our feelings and every kind of prompt and realization that comes to us, apart from outer intrusions. For this we must align energetically with the vibrations of our heart, whose energy is always life-enhancing. It is the feeling of gratitude and joyful vitality. By following our inner guidance, we transform our lives, leaving all fear-based limitations behind. Our personal lives become miraculous.

As we begin this process, we must transcend our ego-consciousness, which exists entirely in the realm of limitation. It is limited to thoughts and feelings, and it does not have true knowing and understanding. Only our heart can give these to us. We can be aware of them in every moment. Although our heart is much more powerful than our brain, its power is restrained for us until we realize what it is. It is not just a physical organ. It is an etheric connection with universal consciousness. In order to realize this, we must align ourselves with the vibratory spectrum of our heart, so that we can become aware of its essence, which is also our essence, beyond our persona.

Living in Deepest Gratitude and Joy

Because we are incarnated in a realm of duality, we have a choice of living in two worlds, one of chaos and destruction and the other in great love and compassion. The world of negativity threatens us in many ways and is aggressively presented to us in the media and through military action. The world of love is quiet and receptive, but powerful. The world of domination and control is artificially contrived and cannot exist without our life force, which we have given it by engaging with it in our attention and realization. The world of positivity is natural and is supported by divine energy, which we identify as unconditional love and light. It is the source of our living consciousness and is the most powerful force in existence. Negativity seeks to destroy it, but cannot without our help, which we have given it by aligning with fear and anger. As we withdraw our support by focusing on gratitude and compassion, we become exempt from its influence, as we elevate our realization into a higher dimension of vibrations.

To our ego-consciousness, this is impossible, because our ego does not understand the power of our realization. The ego wants to fight for its life. For the ego, unconditional love does not exist. If we do not fight, how do we actually realize our greatest life expression? Because humanity is being elevated above duality by the rising energies of our cosmic environment, it happens when we raise our vibrations and align with the energy of our heart. We can direct our thoughts and emotions in ways that are life-enhancing in every way for all beings. Without our attention and alignment, parasites begin to miraculously disappear from our experience. When we are beyond doubt, this is the power of unconditional love. For better or for worse, our energetic radiance attracts resonating energetic patterns that result in our experiences.

By living in gratitude and joy, which we can choose to feel, and residing in a state of serenity, we can become aware of our eternal presence of awareness. We are no longer a threat to anyone and can no longer be threatened, because we are aligned with a realm beyond polarity. Transforming our lives, our realization opens to a higher dimension of living. Our experiences come into resonance with our state of Being, because we are created to be the creators.

Our reality is what we realize it is. We interact with the energetic patterns in the quantum field of all potentialities, and we can draw into manifestation any scenario that we choose by focusing on the quality of its vibrations. When we become aware of the resonance of our heart, we can choose to bring its energy into our experience. We can fill our awareness with gratitude, joy and vitality. This is a personal choice that we all make every moment, knowingly or not. We can live in the radiance of our heart and transform our lives, while we also transform the energy all around us.

Developing Vibrational Sensitivity

We are aware when we feel emotional warmth or coldness or nothing at all. We equate warmth with life-enhancement and coldness with life-diminishment. We can feel these energies in our environment, and we can create them within and radiate them from within our being. Energetically, warmth is positive and coldness is negatively polarized. We feel warmth as love, joy and goodness, and coldness as fear, anger and depression. Our feelings tell us what kind of energy we are focusing on. Always we have the choice of our focus, even in emergencies. It is emergencies that can teach us lessons that we have not been willing to learn. If we let these opportunities pass without serious introspection, we will still need to deal with the issue that is being challenged.

Our guides can give us motivations to move more completely into the light. When we are open and receptive to the deepest knowing of our intuition, and we pay attention to it without any expectations, we can live easily in compassion and joy beyond our limitations. We can open ourselves to understanding the real nature of our Earth-human lives as learning experiences. We are learning to control our focus of attention and realize our infinite Being.

Since we live in a realm of duality, we have aligned with many negative energetic patterns, while not being able to remember or realize our infinite Being. We must awaken ourselves, because all of the limiting beliefs about ourselves are self-imposed. The empirical spectrum of duality that we have locked our awareness within has no real borders. It is all created in the consciousness that we all are, and we can open ourselves to infinite awareness with infinite creative ability. Until we breach the borders of our human awareness, we can live only in ego-consciousness, which has no higher guidance and cannot be aware of our unlimited Self. This takes us to developing vibrational sensitivity.

If we are looking for guidance, it comes to us in ways that each of us can understand, and we can always ask our deeper Self for help. In that moment, our awareness opens to realization as much as we allow. We feel it, and we know it as real. Then it is up to us to follow the vibrations of our guidance and adjust our state of being to resonate with it. We can use our imagination and emotions creatively to come into alignment with the heart of our Being. It is a doorway to Self-Realization.

Aligning with Our Infinite Being

Living in an ocean of life, we are designed to contribute to its enhancement. Enjoyment of life is our natural state of being. In order to deepen our understanding and compassion, we chose to engage realistically with the realm of duality, which we could not take seriously in our natural state. We had to create a separate space in our consciousness, by imposing limiting beliefs upon our human expression. We still live in our entire consciousness, the consciousness of the Creator of all. When we so desire, we can open our awareness to our natural expansiveness by intentionally resolving our limiting beliefs about ourselves through our intuitive knowing. Our human self is an expression of our greater multidimensional Self, whom we do not know in our ego-consciousness.

It is our intuitive awareness that gives us the ability to know our true Self. As ego-conscious humans, we have allowed ourselves to be so distracted by the stimulating energy patterns of the world around us, that we have been unaware of our intuitive guidance. Because we harbor limiting, negative energy patterns deep within, we can, by serious introspection, learn where we hide them. They are attachments to parasitic processes that sap our life force and keep us in fear and dis-ease. Once we realize their presence and their purpose, we have the ability to release them and retrieve our vitality and greater creative power.

Although our attachments to limiting beliefs about ourselves have been necessary for us to live in the limited compartment of consciousness of the realm of duality, we can realize what they are through our intuition. Our intuitive knowing is our way to realizing our expansiveness. There are many ways we can enhance our awareness of it. Being calm in nature while contemplating the energies around us and opening our awareness to the depth of Being in everything. We can learn a deep meditation technique, which can carry us into alignment with the energy of the heart of our Being. We can sing from the energy of our heart, because we can feel it. Inspiring music can help us to elevate our vibrations and open our awareness to a realm of heart-felt energy.

Once we are aware that we can know a higher dimension of energetics, we can intuitively realize its reality. When we accept this, we can live in universal consciousness. It's a matter of developing sensitivity and confidence in our intuitive knowing, which comes from the consciousness that creates us and gives us our life force. It is unlimited in what it can do for us through the energy of our heart, and it is all we need to live abundant, fulfilling lives in service to the creation and enhancement of all life.

Entering a State of Intuitive Knowing

It is important to recognize how we feel in every situation. When we are paying attention outwardly to reports, events, analyses and predictions, we can notice what feelings they elicit in us. If there is any negativity, we have the choice of aligning with it or transforming it in our awareness. Because we are creators with our own vibrations, we are the transformers of any energetic patterns that come into our awareness. If we feel that we are weak and subject to the control of others, we create these conditions for ourselves. Our condition is entirely our own creation, either by acceptance or initiative. What we feel about our condition, and how we believe we are, determines our experience.

In this world we are the absolute creators of the vibratory quality of our personal experiences, conditioned by our beliefs about ourselves. Because of the relentless training and conditioning that we have subjected ourselves to, we are greatly challenged to realize that we have been entranced in a spectrum of consciousness that is self-contained. We can expand our awareness beyond negative vibrations whenever we choose. We must take control of our attention and decide how we want to feel.

To extract ourselves from the realm of duality, we must recognize the limitations we have imposed upon ourselves. They are all based in fear of some kind, including termination of our consciousness. We have been unable to recognize the absurdity of this situation, but if we focus on any of our limiting beliefs, we will realize that they consist of nothing except what we have imagined as real. We have chosen to subject ourselves to the belief that we are without higher guidance, and that we cannot be aware of what is beyond what we already believe we know.

There is a step that we can choose to take, that is beyond our human belief system. It is the step to resolve our fears on every level by opening ourselves to our own inner guidance, which we can only be aware of when we are in the space of gratitude, compassion and joy. This is our natural state of being. When we stop restricting ourselves from our heart-felt convictions and inspiration, we can become aware of our inner treasure.

Everything we ever wanted to feel deeply and to know comes to us in complete awareness, with as much depth as we can accept. As we learn to calm our ego-consciousness and take control of our attention and emotional alignment, we must resolve our limiting beliefs, because they will not allow us to realize our eternal presence of infinite awareness. The process of realizing our true essence and infinitely creative nature happens synchronously, when we are motivated to open ourselves to it and want to know the extent of our Being. We can develop acute inner awareness of the energy of the heart of our Being and absolute confidence in our mastery of all aspects of our lives as humans.

Accepting, Understanding and Loving Everyone

When we begin to realize that we deeply want to live meaningful and fulfilling lives with others who feel the same way, we have many questions, because our ego-consciousness continues to intrude with its struggle for survival in the face of imagined suffering and termination. If we are attentive to what happens in our consciousness, can feel the energy that we align with in every moment. We know the difference between positive and negative energy, between life-enhancing and life-diminishing energies.

When we are staring into the face of an encounter that challenges us with ego-threatening energies, we may not want to be compassionate and loving. When threatened, the ego wants to engage and resist the threat with a desire for revenge and justice. This entire scenario is negatively polarized and cannot be resolved through warfare or retribution. Negativity cannot be corrected by negativity. To transform ourselves from involvement with negativity, we must change our focus to the energy of our heart.

Whenever we feel threatened or diminished in any way, we are in the grip of fear and negativity. If we react with anger or submission, we are aligning with the spectrum of negativity and creating more of the same in our personal experience. All of this is happening in our own consciousness and being expressed through our electromagnetic radiance, which magnetically interacts with compatible energetic patterns, creating experiences for us.

Everything in the universe is balanced. For every negative, there is a positive. We have the freedom to choose to experience either one. The consequence of our decision is what happens in the quality of our lives afterward. If we want lives that we are truly thankful for, we must begin by being in gratitude in every moment, regardless of whether our ego-consciousness imagines threats. Being loving, compassionate, thankful and joyful works best when we are clearly aware of our inner guidance.

By aligning with the conscious life force flowing through our heart, we can inhabit an energetic dimension that is different from the realm of duality, in which negativity exists. We can choose to be in gratitude, which modulates the energies around us to attract experiences that we want to have. When we accept all energetic patterns of positive and negative polarities, we can become non-judgmental and fearless of everything, knowing that we control the quality of every moment by our own vibratory level. This can be a constant intentional choice.

Because everything is part of universal consciousness, nothing beyond our own awareness has any effect on our essence. Whether we realize it or not, we are the masters of our lives in every moment. This is how we are designed and how we operate. If we can understand this, we can become Self-Realized.

Developing Greater Realization

With all of the powerful incoming light and consciousness-raising energies that we are receiving, we are being given opportunities to transcend the perspectives and beliefs that have suppressed our expansiveness and joy. The limitations that we have allowed ourselves to be enclosed in are dissolving under the influx of enveloping love-light. We have a powerful opportunity to realize our own creative potential in alignment with the energy in the heart of our Being. By opening ourselves in gratitude and joy to the most wonderful experiences we can imagine and feel that we are participating in, we can create the lives we truly want.

Beyond our own essence, nothing can affect us. We are sovereign beings of eternal conscious awareness, endowed with the consciousness of the Creator, limited only by our own choices. Because of the energetic fixations we have subjected ourselves to, in order to participate fully in the human experience, we are greatly challenged to realize our personal Self-identity. We have cleverly hidden our true infinite awareness from ourselves, but we have kept one aspect of our Being that can guide us to awaken from the hypnotic trance of empirical duality and to realize who we really are. This guidance is not intrusive and requires our intentional attention.

If we can release our personal dramas and just relax into a state of gratitude and joy, we can begin to open ourselves to a realm of inspiration and limitless awareness within our own consciousness. This is our connection with universal consciousness and infinite creative ability beyond the density of our incarnated presence. In the depth of our consciousness, we have a presence of awareness beyond the limitations of time and space, as well as the ability to project our vibrations into the quantum field for manifestation in our current lives.

If we intentionally keep ourselves aligned with unconditional love, as much as we allow and as much as we can imagine, we identify transcendent vibrational patterns that we can more easily find repeatedly. We can keep expanding our imagination until we can realize unlimited awareness. This is a natural process, and it happens when we want it to in the deepest way and with complete confidence. This is what we are designed to be able to do as infinitely powerful creators. In transcending ego-consciousness, we can trust ourselves implicitly to be clear and intentionally to embody and project our own positive and life-enhancing vibrations.

We Are Being guided to What We Love the Most

We live in a most benevolent time of elevating energies of love and enjoyment of one another and of our environment. Our Sun is brightening along with us, as is Gaia, our Earth Spirit. This is clear on the etheric and causal planes of vibration and is now beginning to manifest on the material empirical world. Because of the current exact alignment of our Sun and Earth with the galactic center, we are receiving direct energetic flow from our Central Sun. The Earth has settled into her magnetic center and is inviting us to do so with our own magnetic center in the heart of our Being. This is the energy that we are being drawn into, and it is the natural vibratory spectrum of our multidimensional consciousness, with awareness far deeper and beyond the spectrum of time/space.

As we begin to awaken to the greatness of who we are, our beliefs in smallness and mortal life begin to dissolve and become unbelievable in our expanding awareness. We do not need to pay attention with our life force to any kind of negativity or limitation. Each limit to our awareness on every level must come under examination, in a search for its true basis. Why do we have our limitations? They keep our realization within the parameters needed to create the dualistic empirical world as real for us. Without our limitations, we can realize that we are our infinite presence of awareness. Our consciousness is part of universal consciousness, the consciousness of the Creator of our universe and beyond. It is beyond our limited imagination, but it is something that we innately know, because it is who we are beyond our limiting beliefs about ourselves.

We can realize that our heart is the center of our Being, and we know that its vibrations are many times more powerful than the vibrations of our brains, but we have reversed our attention from our intuitive heart to our ego-mind, which works through our brain. The mind has physical limitations that the energy of the heart transcends. The heart does not think, it just knows, because it is our channel to universal consciousness. The ego mind cannot imagine the greatness of who we are and it must be transcended for us to have Self Realization. When we imagine our heart's desires, people whose vibrations we love come into our experiences, and we can live where we are most attracted. This happens naturally, when we allow it in complete confidence and gratitude.

Through the brightest beings in our galaxy and beyond, there is guidance that is flowing to us. It comes through the quantum field in the form of celestial and solar bursts of positive, high-frequency vibrations of photons, who are quantum beings of light. Surrounding and interpenetrating us, these gamma rays are elevating the conscious awareness and vitality in every cell of our bodies and all aspects of our self-awareness, if we allow it. We are being enveloped in attractive energy that is enhancing our own heart energy and offering us infinite and unconditional love and joy, filling our awareness and our life experiences with abundance, freedom and realization of our eternal presence of awareness with infinite creative ability.

3.

Resolving the Limitations of Our Consciousness

Calibrating the Intensity of Our Lives

Through aligning with positive, high-vibratory energy, we can expand our conscious awareness into a higher dimension of energetics beyond duality. As we become freer from attachments to our limitations, our energetic presence grows in intensity, as does everything we encounter, which is actually a reflection of our own level of energetics. Everything that happens to us is us creating the energetic patterns that become our experiences. We do it with our feelings and opinions about ourselves. Everything that has some alignment with fear can be made to disappear from lack of our life force. Our alignment can be toward positive, limitless awareness with former attachments.

Being free of emotional attachments is like being free of attachment to our physical presence. This is impossible for the ego, which is comprised of our attachments. Only with strong intention can we transcend this limitation. It must be done in love and compassion, guided by our intuitive knowing. Our body consciousness is part of us, but we are much more. How much more is our choice. Once we are completely aligned with the energetic level of our heart, our potential is unlimited, and we are truly free to be whomever and wherever we may wish to be.

We probably won't want to be within a realm of limited consciousness. This one has been sufficient, thank you. But since we're still here, let's get on with it. Now that we know the energetic level of where we're going, we can practice being in a higher state that we intentionally create with our imagination and emotions. We can be open and receptive to these energies, and we can develop confidence in being positive always. Being attached to outcomes is irrelevant. They will transpire according to the energy of our state of being, how we feel about ourselves and what beliefs we hold about our limitations. We do not need to interfere with this process, although we're free to do so.

Since our conscious state of being reflects back to us the quality of our own energetic alignment, we are always creating the opposition to our intentions in order to keep the vibrations in universal consciousness balanced for us. Any limitations in our consciousness get reflected back to us as encounters in our lives. As we resolve our limitations, we become free to be unlimited, and nothing is reflected back to us, unless we want to welcome it.

Making Peace within Ourselves

Peace among humanity is an innate desire, and it can be present in our inner experiences. It can be achieved by alignment with our subconscious innate self, who lives in positive vibrations, except when we inject negativity. This results in defects in our physical bodies, because it interferes with our natural vibrations. We have free choice to be able to do this. We can diminish our life force consciously and subconsciously, even if we aren't aware that we're doing it. We have deeply-set negative beliefs about ourselves, which we created in order to have a limiting experience in duality. These energies were unknown to us apart from this realistic empirical experience.

If we can be open to the idea that we are multi-dimensional and unlimited in our true Self, we can begin to experience our consciousness expanding. Freeing ourselves from our limiting beliefs can happen with compassion, forgiveness, gratitude and love. Without fear, our limiting beliefs cannot exist. They have no substance apart from our recognition and belief. Letting go of all fear and changing our perspective to love and joy is the expression of changing polarity from negative to positive.

From within duality, we can hardly imagine that there is so much more. From our technological instruments, we know that everything and every being has an expression of electromagnetic energy. Quantum physics has shown us that our recognition of energy patterns of entities results in the empirical appearance of the entity. Photons, for example, change from waves to visible light, when we recognize them. What is real for us, is what we recognize and believe is real. We have the choice of being limited or unlimited in every moment. It depends upon what we believe is real for us.

Energetically we can understand that maintaining a positive polarity results in positive interactions with other energy patterns and beings. When we realize this, our perceptions can change as well, and we begin to realize the light in everyone and interact with it in mutually enjoyable ways. There is another reality right here with us now, but in a higher dimension of energy, which we are capable of living in.

There is a consciousness at the heart of our Being that constantly enlivens us with the life force of the consciousness of the Creator. We are all the same infinitely-conscious Beings, fractals of the Creator, endowed with unconditional love and unlimited abilities. We only need to recognize our Selves beyond our self-imposed limitations to know who we truly are.

The Importance of Self-Trust

Much of our conscious expansion depends upon our ability to trust ourselves. In order to be truly open to higher vibrations, we need to be able to stay aligned with the energy of our heart in every moment. Only by being able to do this can we trust ourselves to stay positive and unencumbered emotionally. This is a challenge, because we have deep attachments to limiting beliefs and cravings that need resolution. We can accomplish this by facing our doubts and fears with compassion and inner knowing. Practicing just being present awareness can be a way to expand our consciousness, because we become able to align with our intuition.

If we trust ourselves, having faced our fears, we can be open to higher consciousness. We can resolve our fears through opening our awareness to the conscious life force coming to us through the heart of our Being. It is beyond time and space, but we can be aware of its presence by how it feels, when we are just present awareness. It is our connection to all conscious beings and has the vibration of unconditional love and joy. This is the creative consciousness of our true Being, and it is life-enhancing in every way.

As we learn to be sensitive to our intuition, we can gain confidence that we are always guided by higher consciousness and always know everything we want to know. This can happen after we clear our traumas with love and compassion. We've gone deep into negative energy and have emotional scars from these experiences, but whenever we choose, and this may require strong intention, we can forgive everyone involved, including ourselves, and move out of those scenes into a positive realm.

Everything we experience is a reflection of our own conscious expressions. Until we resolve our deepest fears, we cannot be emotionally clear and trustworthy to ourselves, because we have been subconsciously creating negative situations, even if we have been intentionally positive in our conscious awareness.

To come into alignment with our subconscious innate being, we can recognize this part of ourselves as our devoted servant, who manages our bodies according to our level of vibratory resonance and enables us to function in the world. By being grateful and appreciative for all that our innate self does for us, we can align our energies with compassion and love. By maintaining this level of vibration, we are becoming expressions of divine light, and we can interface with our innate consciousness to regenerate our bodies.

Our True Natural, Inspired Capabilities

Humans can be very ingenious. Solutions to every problem exist in the quantum field. Through our intuition we can find them and align with them. When we are confronted with a challenge that our ego-conscious mind cannot withstand, we often become aware of our intuitive knowing, and the solution appears to us. With our intuition we can even jump timelines and be in a different aspect of living by changing our polarity and vibratory level, even temporarily. Once we free our awareness beyond duality, we can have so much more fun, because there's no stress, when we live in a world of love.

Before we lived in this realm of duality, we could not imagine negativity or what fear feels like. In our eternal awareness, we could not have polarity. We came into this compartment of consciousness to experience it and to find our way back to realizing our true unlimited Self through our intuitive guidance, which we must seek intentionally to be aware of. We must have a desire to be our true infinite, creator Self. We can find this level of knowing in the energetic life force that flows through our heart and fills our entire Being. It is only positive and supportive of all life-enhancing thoughts and feelings.

We have provided the defects in our bodies and experiences, and in everything that's wrong in our lives, by our limiting beliefs about ourselves. Believing in suffering and death limits us to a level of fear. The ego-consciousness must have fear to exist, because it has no awareness of higher guidance. When we become only positive by paying attention to our intuition, we no longer need the ego to navigate for our survival, because we are led only to positive experiences. Once we realize the invalidity of our limitations through our intuitive knowing, we can resolve them. We are being guided to examine our personal beliefs in terms of polarity and level of vibrations. Since we have created all of our limiting beliefs, we can resolve them, even the ones hidden deeply in our subconscious.

While intending to achieve complete resolution within, we can use our emotions creatively to rise into positive feelings with gratitude. If we can stay in this state of being, even when apparently intimidated, we can realize that only the ego can be intimidated. There is no such feeling in our true Self, but now we have expanded our abilities of knowing the feelings and visions of negativity. It has deepened our compassion and understanding. Now we can expand into a much greater state of Being of unconditional love in awareness of our infinitely creative power.

What Does It Mean to Be True to Our Self?

Being true to ourself depends upon which self we are asking about. We have our ego-consciousness. We have our subconscious self, which expresses itself through the conditions of our bodies. We also have our expanded Self, eternal and infinite, who is always without judgment behind our awareness. The ego is problematic, because it is who we think we are. The clue here is that the ego can live only in the polarity of fear. It consists of our limited beliefs about ourselves.

As we resolve and transcend beyond our limited beliefs, our ego disappears, because we know that we are infinitely powerful creators with our state of being in expanding conscious awareness. We can be radiant throughout our subconscious self, directing it to regenerate our bodies and being confident in our direction, when we are aligned with our intuition.

We can have incredible abilities, once we can be only positive. Until we are, we keep diminishing our life force to accommodate fear of termination of our consciousness. This can happen when we don't know what consciousness is. To cure this, we usually have to face an imminently impending extreme trauma, when suddenly our awareness goes into a state of suspended animation, and we're just going with the flow, not really connected with our bodies. Or we can just learn to meditate deeply and be able to experience conscious awareness beyond the body.

Being beyond body-consciousness communicates to us that we are greater than our ego and draws us into seeking greater awareness. We can become aware of our attachments to our beliefs in limitation about ourselves, and we can examine them in the light of expanding consciousness. As we resolve our beliefs in limitation, our ego weakens, because there's less to fear. Without fear, we can become sensitive to our intuition and learn to align with its vibrations.

Being able to resolve and transcend our attachments to limiting beliefs about ourselves happens when we can align with our intuitive knowing, and we can no longer believe that we are limited. Instead of paying attention to our ego-consciousness, which is always struggling to avoid suffering, we pay attention to our intuitive knowing, which provides everything we need. Once we are sensitive to how it feels in every moment, we can become Self-assured.

It is possible to penetrate our intuition deeper and deeper to reveal more intensity and greater diversity in subtle ways. If we keep seeking to expand our awareness into more positive and higher vibrations, they come into our awareness. When we are open and receptive to them, we can recognize them. When we can believe them to be real, they come into our empirical experience.

The Dawning of Our New Day

When we are seriously on the path to inner knowing of our expanding Self, we can begin each day greeting the Sun with gratitude and opening ourselves to the infinite One that we can know and align with. We can ground ourselves by walking barefoot on the Earth for a while. Listening to the inner sound current brings us into present awareness all around and draws us into a higher vibratory state, in which we can more easily be joyous and expansive.

In this state of being, we can transcend our ego-consciousness and invite our guides and angels to draw us into feeling their presence, as we open to higher consciousness. All conscious beings have a radiant presence, which we may feel. As we live through each day, we can feel the quality of energy that we pay attention to. At the same time, we can remain in an elevated state of joy and gratitude for each experience. Any challenging situations, in which we face strong negative energy, can be transformed by our steadfast alignment with a positive perspective. When we are completely positive, consciously and subconsciously, negative energy cannot reach us, because it is not real for us, and we no longer give it our life force. It is only a memory.

Nothing happens by accident, and everything is patterns of energies, some of which we perceive as our reality. We are able to change the quality of energetics in our presence by our perspective and state of being. It happens in every moment, as we direct or accept our thoughts and feelings. We are not punished or rewarded. We experience the qualities of energy that we pay attention to and align with, both in our empirical experiences and in our imagination.

Devotion to aligning with our inner truth draws us to experiences that enhance this perspective. Our experiences may not make sense to our ego-consciousness, which gradually fades away, as we learn to recognize our intuition on a deep level. Our limiting beliefs fade away as well, because we can become aware of our expanding eternal present awareness.

When we realize that we are limited only by our own beliefs about ourselves, we have the option of resolving them and opening ourselves to our expanding awareness. All of our limiting beliefs can be resolved by our intuitive knowing and cooperative alignment with our subconscious through compassionate understanding.

Opening to Our Intuition

Once we decide to open ourselves to our intuitive knowing, we face the great challenge of transcending our ego consciousness, which we have identified with for our entire lives. Whenever we are criticized or belittled by others, we feel ashamed and weak. When we are praised and rewarded, we feel proud and strong. In reality, these feelings are part of our ego, not our true Self. Intuition comes from our higher consciousness and is always in the background of our awareness.

In great love and compassion, intuitive knowing comes from a Source beyond time and space that knows us intimately and wants the best for us always. With a feeling of expansiveness and calm serenity, it can be accompanied by miracles when needed. When we are open to it and learn to depend upon it, our intuition cannot be defeated by circumstances. It provides the experiences that we most need and deeply want to fulfill our life plan.

In the world of ego-consciousness, intuitive knowing is not believable, because it has no basis in empirical reality, which is the only vibratory level on which the ego operates. As long as we believe that we are finite beings, separate from everyone else and from our Creator, we cannot accept our intuitive knowing, which comes from our infinite unity with all conscious beings within universal consciousness. Our ego-consciousness keeps us limited, because this is what we need to experience the complete spectrum of human experience.

It is only when we are ready to ascend back into our unlimited, infinite Self, that we decide to transcend the ego. We can begin this process by praying within ourselves and asking for our inner guidance in every situation. We can imagine and feel being divinely inspired to embody the positive, high-frequency vibrations of love, gratitude and joy as much as possible. As we open our awareness to these feelings, we begin to know how to be in every situation.

If we continue to imagine that we are each an aspect of the divine One, we can feel our connections with one another and other beings in ways that support and enhance all of life. We all share the same Being in our own unique ways, and we all contribute to the expressions of the One in universal consciousness. As we practice our inner alignment with the divine, our knowing becomes unlimited wherever we choose to focus our attention, and our lives become expressions of love, abundance and freedom, regardless of what may be happening in the world around us.

Our Creative Presence

When we contemplate the idea that we are the creators of our life experiences, we may recognize that this is more than a mental process. Our entire being is creative to the extent that it contributes to our state of being. How we feel about ourselves is also a state of knowing. If we realize that we have defects here and there in our bodies and personalities and subconscious drives, we can also realize that our creative process has a complex shield of limitations that we have to resolve, if we want to be clear in our consciousness.

When we are clear, we can create what we desire without constraints. We can be fearless and at ease. We can fully confront any energetic patterns that appear in every moment.

There are destined experiences that we cannot control. We can, however, control our reactions to them, through our awareness of the quality of their polarity and vibrations. We can be aware of these things quickly in every situation. If we choose or are required to engage with the energies that are brought to us, we can enhance them or diminish them with our thoughts and feelings, or we can just observe. Every experience has an energetic quality, and we may either align with its vibratory resonance or transform any negative into positive. If this cannot be accomplished, we can dissolve it from our experience by transferring our attention to a positive level of compassion and love.

How we think and feel about ourselves, consciously and subconsciously, determines our state of being. This is our energetic signature that we radiate out into the quantum field that envelops us. Here we repel and attract energetic patterns that disappear or manifest into our personal experience. We have electrical and magnetic polarity in our thoughts and emotions. Humans mostly have a dualistic polarity of both negative and positive, but primarily we are negative, because we are fearful at some level.

To be clear, we need to know that we are immortal. Our present awareness is always with us. Many people who have died and come back confirm that our awareness expands greatly beyond the body and is always present. Deep meditation, in whatever form works for us, can help us to transcend our physical limitations. Slow, deep, rhythmic breathing also calms the mind and allows for transcendence.

Being quiet and aware in nature helps us to align with the resonance of the Spirit of the Earth. This is a positive resonance of unconditional love in the enhancement of all life. As we become more sensitive and receptive to the energies of nature, our awareness can open to the realm of nature spirits and other beings living in resonance with Gaia. We can use our creative presence to help all of the beings in nature to enjoy greater love and joy.

Transforming Our Experiences by Transforming Ourselves

Who we are is our eternal present awareness. Without infringement on others, we can focus our awareness on anyone and anything we desire, and we can just be present and aware, without evaluation or judgment, and at the vibratory level of gratitude. With this alignment, we can neutralize or transform any negative energetic pattern that wants our life force. It happens through the realization of unconditional love and universally-conscious personal awareness.

To be able to open ourselves to this extent, we can recognize the true validity of all of our beliefs about ourselves. If they are intended to enhance all of life, they are coming from our own intuitive knowing and are an expression of universal consciousness. If they are based on some expression of fear, they are coming from our ego-consciousness and have no essence of their own.

We can begin to become aware that, as humans, we are living in a limited compartment of consciousness, veiled off by our limiting beliefs about ourselves. We self-limit ourselves by having doubt that we have unlimited creative ability. Doubt disables our creative expressions. This is part of living in duality, and is why people have difficulty manifesting things. They have some doubt or fear, some limiting beliefs.

Once we have thoroughly examined our beliefs, we can know the quality of energy they are based on and can decide what to do with them. Our choices always are to align with them, to transform them or to dissolve them by withdrawing our attention and belief in their reality. These are our abilities, which we exercise through our thoughts and emotions. Our personal beliefs have their existence only by our intention to be limited, which we have needed, in order to participate fully in the human experience in dualistic energetics.

By following our intuition, we can enable ourselves to accept all energetic patterns that we encounter. In this perspective, there are no mysteries in our experiences. It's as if we're in a play, and many scenes are pre-scripted. The important choices we face in these situations are regarding our own state of being. If we can be in an emotional level of compassion, forgiveness, gratitude and joy, which we can practice creating, we can align ourselves with this energy level. The longer we can hold this perspective, the more we can express greater positive creative energy. This results in an expansion of awareness and appreciation of higher guidance within our own knowing.

Enchanted Being

Perhaps we can begin to imagine experiencing the most powerfully alluring personal fulfillment in every way. This varies for everyone, but there is one dominant vibration, and that is of being unlimited. We can be unlimited in experiencing eternal love, joy, abundance and infinite creative ability. These are the feelings and abilities of the universal Creator, of whom we are fractals in our Self-Awareness.

Together with all of humanity, we have created the empirical matrix of our beliefs. We each have ego-consciousness, created out of the desire for limitation, which we either support or accede to. All of these limitations are self-imposed and can be self-released. They are set deeply in our psyche and can be resolved through transcendence into expanding present awareness.

Because we got stuck in negative feedback loops of suffering and death of the body, we developed deep fears. They could develop only because we were unaware of our unlimited nature. Once our awareness opens beyond time and space, our limitations become unbelievable, and we can drop them. This can transform our human lives, which become an obvious play of mostly-scripted characters and scenes for our fun and learning, including all the emotional drama. We have the conscious choice of influencing the script by our creative responses in our experiences, and in our predominant polarity and vibratory level. Everything is designed to move toward compatibility.

By holding our focus mostly with gratitude in every situation, we can experience mostly situations that elicit that perspective. The more we can stay in gratitude in every moment, the more our lives improve in every way. In gratitude we can live in clear objectivity, in acceptance and forgiveness. By directing the quality of our perspective and our emotional state, we create the quality of our experiences. As our energy radiates into the quantum field enveloping us, compatible energetic patterns are attracted to us. As we recognize them and believe that they are real, we experience them in our lives.

We are multidimensional without realizing it. Our thoughts and emotions have no counterpart in the empirical world. They are strictly aspects of our conscious awareness, and they have not been entirely quarantined within our limited human consciousness. We can have wonderful visionary and imaginary creations, in which we can feel ourselves living. These kinds of endeavors influence the destiny of our encounters in life.

By using our creative abilities in our attention, we can transcend any limitations and challenges. If we can be intentionally positive as much as possible, our lives will fill mostly with love and joy. By practicing transcendence, we develop a strong perspective of being present in eternal Self-Awareness. This is the beginning of mastery, of creating miracles and of living in unity of awareness with all conscious beings.

Releasing the Limits of Our Awareness

360-degree spherical awareness without bounds and beyond time and space. Is this an attribute of our expanding consciousness? In a way, we can begin to imagine it. Being fractals of universal consciousness, we have this ability. It's available when we transcend our limitations. Once we withdraw our attention from our beliefs in what is real, we enter the unknown, at least to us. We are expanding our awareness.

We need guidance, and we can turn our attention to our feelings, which we can become acutely aware of with our increasing intentionality. Our intuitive knowing brings everything within the scope of our potential awareness. Everything has an energy pattern, and our intuition knows this and communicates it through our emotions, often very subtly, but in a richness that we can understand.

Even if our conscious awareness is not that expanded, our greater Self is, and it communicates to us everything we need and want, subject to the limitations we impose upon our awareness and our perspective. We determine what is true for ourselves by our beliefs. As we become more aware, accepting, forgiving and transforming ourselves, we transcend our limitations in compassion and gratitude.

Currently, many new positive energies are enveloping us and causing much instability, as the old negative energies of the human world break down and dissolve. The negative feelings and entrapments are becoming incompatible with the rising vibrations of the Earth and of humanity. This creates conflict and chaos in society and calamities in nature, while in the background is arising a new world of love, beauty, abundance and freedom. We are in the transition zone and just need to keep ourselves positive, while we work through our attachments to limitation and move into a higher dimension of living.

When we realize that our true Self has no limitations, we can align with the unconditional love of universal consciousness. The compartment of consciousness that we have kept ourselves in for eons is no longer needed. We can become truly sovereign and free, fully present in eternal, infinite awareness, confident in our unlimited creative power.

Understanding and Transforming Our Limitations

Although we have been taught that we are imperfect in so many ways, this understanding is actually a matter of choice on our part. We have absolute control over the quality of our being by our energetic orientation and vibratory level. It is how we imagine ourselves to be. We know the quality of our consciousness by how we feel about ourselves, and we have a wide range of options available to us.

When we are young and impressionable, just learning to adjust to life as humans on this planet, we have little defense against accepting limiting beliefs that enslave us to those who desire to control us, but the galactic energies are now becoming more positive and elevating in every way. We are being prompted to question our limiting beliefs about ourselves and to choose freedom and fulfillment, instead living under duress.

Our human perspective is the belief and appearance that we are separate individuals, conceived and birthed to be vulnerable to suffering and termination; yet we have achieved a technological capability that has enabled us to determine that this belief and appearance are tricks that we play on ourselves. When we examine our reality in the most minute ways, down to subatomic waves and particles, we find that our material world consists entirely of electromagnetic waves of energy with patterns of polarity, frequencies and amplitude. What seems solid is nothing more than our conscious interpretation of these energetic patterns. We can perceive them with our senses, feel them with our emotions and imagine them with our thoughts.

What has become clear for us is that we are not physical bodies; rather, we are conscious entities in a sea of many other conscious entities, all arising within a greater consciousness that provides the life force for us all to express the essential qualities of the Being whose consciousness we live within. Those qualities are our personal awareness and our ability to know the essence of our Creator and to create experiences for the greater enjoyment and enhancement of all conscious beings.

With our technology, we have determined that everything that exists has conscious awareness of its own being. As fractals of the Creator, we have the potential awareness of the entire consciousness of everyone and everything. Many of us have experienced this and know it is true. It is only our limiting beliefs about ourselves that maintain a veil of unknowing in our awareness. These beliefs are self-imposed, as a result of our early training and programming, making it possible for us to resolve and transcend them through our awareness of the presence of the Creator in our own intuition. By practicing what we can imagine and feel in ourselves as the greatest love and joy of the essence of our Creator, we can sensitize ourselves to our intuitive feelings and awareness.

Releasing the Source of Our Limitations

Whenever we feel anxiety approaching, we can intentionally pay attention to our high-frequency inner sound current. If the sound originates in the heart of our Being, it is always present, as is our intuition. It wants our attention. Our inner sound stabilizes us emotionally, allowing us to minimize the impact of negativity that we may confront. We can feel the negativity, but we can maintain our alignment with the unconditional love radiating through our intuition.

We can begin by imagining and feeling greater and greater positivity. By using our intentional will-power, we can maintain a focus on our inner sound current and our intuitive knowing and feeling. This is training for our ego-consciousness to obey our encouraging directive to pay attention to our inner promptings, as well as the symbolism of the experiences in our daily lives. These are portions of our inner guidance from higher consciousness. Once we commit ourselves to full awareness of our eternal, infinite Self, we enable our awareness to expand into greater positivity in deeper compassion, love and gratitude.

By practicing being positive always, we can learn to be absolutely confident in our state of Being. In this perspective, only positivity can affect us, because we're in alignment with positive polarity, unless we align ourselves with negativity by resisting or embracing it. In either case we would align with anger and fear. When we are vibrating at the frequency of love and compassion, we cannot be challenged by negativity.

If we can transcend our rational, empirical, dualistic beliefs about ourselves, we can enter a different dimension of consciousness, where only positivity exists. Here we can have complete confidence in our creative ability, ensuring that we have transcended self-sabotage of our intentions. We can achieve this perspective by repetitively examining, resolving and training our subconscious, until we are clear throughout our conscious awareness.

All forward progress depends upon resolving our personal limitations, which are reflected in our experiences, and which have an emotional presence. We can release our attachment to all energetic patterns based in fear. These exist only because of our attachment to them and have no other source of expression. Without limitations lurking in the background of our awareness, we are free to expand as far out as we desire toward our infinite awareness and unlimited creative power beyond time and space, as well as within the empirical world.

Opening Our Conscious Realization

We are so much more than we have believed. Through early telepathic experiences, training by parents and schools and our local society, we have come to believe that we are localized, physical beings. Some also believe that we have a soul that is beyond our perceptive ability. There is a growing number who have realized that we are living in a realistic simulation of experiences that is an expression of greater consciousness. Some believe that this consciousness is artificial in the sense that is has no inherent life force of its own. This is for each of us to recognize. We are its creators, imagining it from ego-consciousness and endowing it with our attention and energetic alignment, apart from the essence of our heart. It is a mental construct that we have chosen to believe is real, so it is real for us. But it operates within a limited spectrum of energetic duality within the empirical experience.

Scientifically and spiritually, we know that there is a universal consciousness that we participate in, but we have established boundaries to our awareness. These boundaries were created by us, so that we could have an authentic human experience. If we decide that we want greater awareness, we can work with our consciousness to resolve these boundaries, by paying attention to our intuitive guidance. We can also work creatively with our subconscious, innate being. Our subconscious is our faithful servant who forms and maintains every cell of our body. It learns repetitive patterns of movement, and it manifests the energetics of our predominant mental and emotional patterns into the enhancement or diminishment of our body. It expresses our own qualities of energy into our physical presence.

Our subconscious knows our intuition and would naturally follow it, but the subconscious is primarily subject to our conscious direction and guidance. We have given this task to our ego-consciousness, which is a creation of our limited beliefs. The ego is unaware of our intuition and must be transcended, as we become sensitive intuitively. Intuition has no boundaries and can be known emotionally, mentally and sometimes through the senses. It works through our entire being, every way that we can be aware. When we learn to align with the energetics of our heart, our subconscious comes into alignment with our conscious awareness, and we become a unified Being, with a positive perspective.

As we align ourselves with our subconscious self in the energy of our heart's intuition, and we have become a unified intuitive Being of unconditional love and compassion, our awareness opens beyond the world of human experience to our infinite, eternal present awareness. We can realize ourselves as infinitely powerful creators in all dimensions. As fractals of universal consciousness, we are connected to our essence through our intuition and are designed to create knowledge, emotions, and experiences beyond the capabilities of ego-consciousness.

Expanding Our Expectations

In the quest for expanded consciousness, we can imagine the most glorious life experiences that we may be capable of accepting. Once we have cleared ourselves of incursions of negative alignment, we become unlimited in our presence of awareness. We can be aligned with loving intent in all encounters, realizing that we are all aspects of One Being with infinite creative power, which is available to us, when we have become mentally and emotionally clear and finely attuned to our intuition. By deciding to let go of our attachments to limitation, we can resolve our false, limiting beliefs about ourselves.

The resolution of our belief in limited, temporal consciousness can occur in various ways. We can search for deep memories of former incarnations in various dimensions, as well as physically- and hypnotically-conscious, out of body experiences. When we decide to move beyond duality and embrace only positive energies, the energetics transform our lives, and we can begin to live in gratitude, love and joy in our state of being and in our encounters.

When we can recognize the limitations of our ego-consciousness, we can resolve them by focused intent with compassion and understanding. We can be thankful that our ego-conscious-self brought our awareness to this moment without higher guidance. Ego does not recognize intuitive knowing, resulting in constant stress, frustration, anger and fear. All of this can be resolved by changing polarity to only positive. Here there is only success, because there is no other option, without returning to duality.

With practice, we can learn to be only positive. By confronting negative beliefs with intuitive knowing, we can train our subconscious to be only positive. Since our entire body is controlled by our subconscious, complete positivity can even result in regeneration of our body. Without our belief in their reality, negative beliefs have no conscious life force supporting them, and they dissolve out of our experience. Thus, we become capable of accepting more wonderful encounters. The realization of expanded awareness is something we achieve by just being present, without attachments, in the moment.

This entire process of realizing the truth about our unlimited Selves, is an individual and intentional endeavor. It's different for everyone. In my writings, I'm just describing my process and what's happening in my awareness. I write in the plural, because we're all the same Being. Because of our telepathic connections, what works for one, potentially works for others.

Transforming Anxiety and Other Manifestations of Fear

Being truly open to our higher guidance implies the ability to focus our attention without extraneous concerns. We must be able either to resolve our limitations or to transcend them, so that they are no longer in our dimension of energetics. In order to transcend our limiting beliefs about ourselves, we can make a leap in consciousness. This can also happen with very intentional use of psychotropic plants and mushrooms, as well as very deep meditation, and other methods of opening our awareness beyond our limitations. We intentionally can change our recognition of reality. We already partially do this with our computer-generated augmented reality headsets. If we take this path, our intention must be powerful enough to dissolve the grip of limitations by no longer believing in their reality. It's an advanced form of professional acting that envelops our awareness. Unless we feel that we can survive all of this, we can just as well proceed on our path in our own way, perhaps with greater awareness.

Suppose that we decide not to completely repolarize ourselves, but we want to continue our process of resolving our negative attachments, including our deepest pain, while we are heading toward recognizing our infinite present awareness. If we align ourselves with the energies of our heart and open our awareness to positivity in as many moments as possible, we can learn to adjust our vibrations to more fulfilling levels of our perception.

When we feel strong anxiety and stress, we have deep, unrecognized fear, which we can have only by believing in our finiteness and mortality. This is what we must realize and confront energetically. Finiteness and mortality are life-diminishing and fearful. Knowing this, we can learn how to interact with our intuition and eventually know the eternal, limitless presence of awareness that we are. Intuitively we have access to universal consciousness, enabling us to know everything we want to know.

Within the limitations we have placed on our awareness, we are aware of very little of our intuition. When we want to expand our intuitive awareness, we can focus on the energies that we like the most and feel a warmth toward. This is the guidance of our heart, and we can follow it in our awareness. As we follow our warmth, we are led toward limitless intuitive knowing and feeling. Eventually we feel and know that we are infinite awareness, and we live in a dimension of unconditional love and abundance.

Transforming Our Limitations

A most difficult part of expanding our conscious awareness is releasing our limiting beliefs about ourselves. They are based on the shame and fear of negativity, and they comprise the nature of our ego-consciousness. When we resolve our limitations, we also resolve our ego, leading us to the unlimited Self-Awareness that envelops the cosmos. We are infinite Being, fractals of the Creator of all. Ego cannot imagine who we are. All it knows is limitation. Through our intuition, however, we can know our true Self, realizing our present awareness beyond space and time. We can develop intuitive sensitivity in opening to higher consciousness.

Intuition is always life-enhancing. It has only positive polarity, and its Source knows much more about us than our conscious minds do. In our human desire to be separate individuals with a private consciousness, we have kept ourselves unaware of the Source of our conscious life force, which is universal Being and which enlivens us in every moment. We live in the consciousness of the Creator and are enveloped in the unconditional love energy that enlivens everyone. In our human life, however, we have developed negative perspectives that disallow us from paying attention to our divine knowing.

Our hypnotic trance of negativity can be awakened by intention. When we intend to be aware of the energy of our heart, we open ourselves to the quality of its presence. This is the beginning of expanding awareness. If we can be in a serene place in nature and just feel the energy when we focus on our heart, we can feel the enhancing energy of Gaia. If we are open to the highest vibrations, we can begin to be aware that every cell in our body is luminous. We have inner light. Our cells constantly emit photons, which are quanta of light. Our presence is radiant.

If we examine the nature of consciousness, we find that it is not localized in us. It is universal, shared by all conscious beings, including the consciousness that creates everything. We are telepathic with all creatures and things. Consciousness creates and sustains everyone and everything. It is the force of life and awareness. It expresses itself in the unified quantum field of all potentialities. In our mental and emotional creative essence, we align with energetic patterns in the quantum field, radiating our thoughts and emotions into the enveloping plasma for manifestation in our experiences.

When we know and follow all of the subtleties of our intuition, we can transcend the realm of duality and live intentionally in the realm of abundance, love and freedom. We can learn to express the desires of our heart in gratitude, compassion and joy, and we can expect that life will reflect the quality of energy that we consciously align and resonate with. Without limiting beliefs, we are free to be our true, unlimited Self-Awareness, even as humans.

The Physics of Self-Transcendence

To be truly and completely free is our destiny, and it is possible when we release all of our attachments to our beliefs in our personal limitations. As long as we believe that we can be threatened, mugged, enslaved and murdered, we cannot be free. If we believe that these things are possible for us, we are aligned with this level of energetics, and we believe that we are separated from the Source of our life. Even if we recognize that this is impossible, deep within we hold onto irrational fear that we have inherited from terrible experiences in antiquity, as well as our social programming. If we can recognize this, we can retrain ourselves by intentionally calling for our Source Being to draw us into awareness of our unconditionally-loving higher Self. When we do this, if we are calm and open, before we can even take a breath, we are flooded with the energy of divine love enfolding us as a tingling feeling throughout our body, alerting us to our divine Presence.

If we can relax into awareness of our eternal presence, we can begin to release the limiting beliefs that have enclosed us within a compartment of universal consciousness that we recognize as the human world of dualism. It is only our limiting beliefs about ourselves that hold us in negative experiences that result in pain. suffering and fear. Nothing beyond our own beliefs has any power over us. Within the hypnotic trance of human experience, we cannot realize this as true. Our ego consciousness cannot expand into infinite Being. To realize this, we must transcend our limitations, which manifest as our ego-consciousness.

There is an aspect of each of us that knows our truth beyond the ego. It is our intuition, coming into our awareness through the energy of our heart. This is where we know our eternal Self and our connection with all conscious beings through the consciousness of our Creator. When we seek to be aware of our intuitive knowing, we can come to realize our essence beyond time and space. This happens on a vibratory level.

The empirical world consists of swirling patterns of energy, beginning with the subatomic beings manifesting as photons, protons, neutrons, electrons and their colleagues, all combining into larger, more complex beings, and resulting in the world we recognize and feel. Our bodies consist of trillions of conscious entities, all gathered and directed in alignment with our consciousness. They are all intended to function in divine vitality to the extent that we live in the unconditional love of the consciousness of the Creator. To the extent that we entertain and believe in fear and limitation, we create the dissipation of our life force, resulting in aging, disease and physical death. None of this is necessary. We can transcend our limitations by resolving and releasing them intentionally with much practice and self-examination. resulting in realizing our essence in Source Consciousness.

Creating a Wondrous Life

In the essence of our heart, we receive the Creator's conscious life force in unconditional love and universal consciousness. We are free to use and modulate this subtle energy however we choose. It is the energy that creates universes, and in our conscious presence of Being, we have access to all of it. In our desire for the most powerfully impactful experiences, we chose to embody as humans in an empirical world with the possibility of negative experiences, which we have to create ourselves. Negativity is not part of divine creation. It exists for us only by the vibrations created by humanity.

We use our life force to manifest our experiences by how we feel about ourselves. Our entire self-image is self-imposed by the choices we make in aligning the vibrations of our mental and emotional states of being. This is where we create negative experiences for ourselves. All of these depend upon beliefs in our limitations. These are a trap that we have set for our unlimited consciousness to become limited, in order for us the have the most impactful experiences. We want to be even more loving and compassionate, with greater wisdom than we could have realized without these experiences in duality. Although we have now achieved great destructive ability, until recently our limitations have kept us from being excessively dangerous.

All the energies in the universal consciousness that envelops us are drawing us into higher vibrations, along with our planet, solar system and beyond. As we awaken and become trustworthy to live in the enhancement of all of life, we no longer need limitations, and we can understand them and release them. We are all expanding our awareness beyond time and space to infinite, eternal presence of Being and unlimited creativity in unconditional love.

To be aware of our expanded Being and to enjoy living in high vibrations, we can imagine our current circumstances from a perspective of love and joy, while relying on intuitive guidance coming through our heart. There is light, even in very dark beings. We can recognize it and relate with its energy, not aligning with the negative energy that may be present. By not engaging with its energetics, the negative disappears from our lives, because we give it no life force. Consistently intending to interact with all energies that we encounter from a high-vibratory perspective, we can maintain an energetic radiation that attracts high-vibratory experiences. Wonderful experiences of all kinds come to us, because this is the vibratory world that we can live in. By intending to keep our own vibrations positive and high, we can live in abundance, freedom and deepest love, gratitude and joy.

Becoming the Champions of Enlightenment

What does it feel like to release all of our attachment to victimhood? It is perhaps unimaginable for us to realize true freedom from any constraints within time and space, but it is our natural reality. We are our pure presence of awareness, unlimited in every way and able to be any kind of entity in any dimension with any powers we wish to instill. In our current status as humans on Earth, we could not believe it is possible, because of our self-imposed limitations, which are deeply set into our species consciousness. The power of our beliefs in limitation, for as long as we hold them, constrains us within their limits. Our vibratory level attracts circumstances and encounters that resonate at the same level, creating experiences below the level of mastery.

It is very difficult for people who have worked hard for a long time to accumulate something, to then allow themselves to be unattached to their attainment. But it is the attachment that prevents further expansion. As long as we believe in lack of anything, we create its experience in our lives. The challenge is to resolve our beliefs by transforming our fear and doubt about our abilities into confidence and joy. This does not require any special talents on our part. Anyone can do it and become a master of life on Earth and much more, because it is our natural state of being. Our situation is further complicated by our belief that our beliefs are not responsible for our predicaments. We have believed that things just happen to us, and we have no control over what comes into our lives. We have believed in accidents, and so we have them.

What we experience in every moment comes to us as a result of the operations of our consciousness. It is our level of vibration and our perspective. The primary consideration is our polarity, positive or negative. Do we vibrate at the level of love or fear? It has to be one or the other. Love is ultimately confident, free and supportive of everyone. Fear experiences a sense of lack and frustration, while blaming others. Fear is the basis of every problematic situation.

In our true essence, there are no problems. We are infinite Being. We are everyone and everything everywhere. It is all part of our presence of awareness, our Self-Realization. This may be so far beyond our current self-understanding, that we cannot imagine being infinite, but we can take steps to get there, and realizing our limitations is one of them. Then facing them one-by-one with serious scrutiny to understand their value for us is another step. If we can learn that we have the power in our intention to change and transcend our beliefs, we can be successful in dramatically raising our vibrations, continuing to expand our awareness and deepen our experiences of love and joy.

Enhancing Our Human Potential

As we exist on Earth today, we humans have incredibly expansive capabilities waiting for our realization. We have essentially enslaved ourselves in order to experience the reality of living in negative conditions with just enough positive to keep us barely

Chapter 3. Resolving the Limitations of Our Consciousness

functioning. Because we are naturally curious, we wanted to know what we could experience, if we gave up much of our awareness of ourselves and gazed into the diminishment of our self-awareness, while being embodied in convincing physicality. Imagine what we can be without a physical body. There is no way we can completely convince our ego-consciousness that we can exist without our body, because the ego consists of our limited beliefs about ourselves. These have been necessary for the intensity of our experiences, but there comes a time for us, when we wonder about what might be beyond our current abilities.

The only way to find out is to transcend our beliefs in what is real. It is here that quantum physics can be helpful. Physics experiments have shown that we live in an indeterminate world of subtle energies that span frequencies far beyond the awareness ability of our senses. They have also shown that everything has conscious awareness, including all sub-atomic entities, including photons, electrons, protons and neutrons. These are conscious entities with full awareness of their cosmic environment, their pathway of movement and their focus of energetics. They know how to interact for the benefit of all. This is our natural energetic state of being. It's when all of our cells are fully functioning as they are designed in universal consciousness.

Physicists have found that the consciousness of subatomic entities is unlimited in terms of their inherent life processes. Our bodies consist of sub-atomic entities, all having cosmic consciousness. So what about us? Where's our cosmic consciousness? It is present in all of us, waiting for us to open ourselves to its realization. For this to happen, we have to get past our preconceptions and our limiting beliefs. We can use our imagination to pretend that we're living in a realm that we deeply want to experience. We can open our awareness to our intuition by aligning with the energy of our heart.

As our ability to do these things increases, our awareness begins to transcend ego-consciousness. Our understanding of reality begins to transform into a natural positivity, in gratitude, love and joy. In every moment that we're in this state of being, we create an experience of its energies for ourselves. Once we can feel its reality, it becomes real in our experience. We are the creators, and we set out own limits and transcend them.

Realizing the Extent of Our Love and Compassion

When we begin to realize that we may have no real limits to our Being and our consciousness, we can begin to transcend the conditions that we have imposed upon ourselves. We can question everything that holds us within limitation. Do we need to have enemies and allies, or are these self-created to enhance participation in our quest for Self-Realization? Do we actually have needs and wants that are unfulfilled, or are these self-created?

One way to know the answers to these questions is to open ourselves beyond ego-consciousness, which may be a challenge. It could require deep meditation, deep breathwork, a near-death experience, a guided psychoactive experience (such as ayahuasca or psilocybin) or something else that releases us from our normal hypnotic trance in ego-consciousness. Another way similar to deep meditation is to develop extreme introspection, leading to deep sensitivity to our intuitive knowing. This is our most natural approach to enlightenment and is a lasting experience that can continue in every moment.

Once we are connected intuitively to our inner knowing, we do not need anything beyond our own Being, because we envelop everything and everyone in our potential awareness. We participate in the perspective and intent of Creator consciousness in unconditional love and infinite creative power. There are no enemies, and there is no struggle, because we are beyond the reach of threats and intimidation. We are complete in ourselves in the eternal presence of awareness, living in ecstasy and freedom.

This doesn't usually happen instantaneously. It is a life-long process that has now been greatly accelerated by the intensification of the light in our cosmic environment and the rising vibratory frequency of our planet. We are being prompted to transcend our ego-consciousness by opening to love and joy in every aspect of our lives. When we are uncomfortable with someone or with some circumstance, we can look within ourselves for the energy that needs to be adjusted. We can look for the life-enhancing energy that we have been unaware of, and we can ask our guides for help in realizing it, while maintaining our focus on the energy of our heart.

In our own essence we can be aware of an unlimited source of creative power that can elevate us and every circumstance that we are involved in. It is present for us in our realization of it. By learning to control our thoughts and emotions and to project them at will in alignment with the energy of our heart, we can develop our power of realization and open ourselves to a new world of love and abundance.

Living beyond Our Limitations

The most intimate connection with our Creator is the conscious life force that we constantly receive through the heart of our Being. This is our link to universal consciousness as well as the connection with our physical presence. Our ego-consciousness is our own creation, stretching back into our ancestry, and it is our navigator through the realm of duality and limitation. We are our conscious essence, our presence of awareness and self-realization. We can feel our conscious life force flowing throughout our awareness as an inner stimulation that feels filled with vitality and joy. Any negative feeling is actually parasitic to the flow of life force. Negativity depends upon our own creative projection and requires a portion of our life force for its existence. If we give our attention only to positivity, we are in our natural energetic flow, and negativity cannot exist for us. Our attention is in a higher-frequency band of vibration.

The magic of this understanding of energy is that it also translates into the empirical world. For the ego, magical happenings are mysterious, but understood energetically, they are natural manifestations of the quantum world. This is the energetic realm that is hidden behind the empirical world of our physical experience and is the pattern that manifests as our experience. We have access to the design of our lives by the vibratory energy of our mental and emotional abilities, which results in the personal resonance of our energetic signature. Our mental and emotional patterns are our energetic radiance within the quantum field.

Either intentionally or by acquiescence, we control our mental and emotional vibratory alignment. We are energy modulators. By our mental and emotional control in every moment, we are able to focus on the mental and emotional qualities that resonate with the energetic pattern of our heart. This results in life experiences that are magical in love and gratitude.

When we realize that the belief that what we experience causes how we feel and think, is backwards, then we can understand that our emotional and mental states are the cause of our experience, and we are on the path to mastery of our lives. We can learn to recognize when we are operating in limitation and change our perspective to alignment with our infinite presence of awareness.

Just being present in awareness and holding a focus on unconditional love and fullness of Being, as much as we can, is an exercise that will ultimately carry us into realization of our infinite Self. In the realization of our infinite and eternal presence of awareness, we have access to universal consciousness in the essence of the Creator, which is also our essence. In this realization we are completely fulfilled.

Clarifying Our Consciousness

As we continue along the path to inner knowing and expanding into universal consciousness, we can seek energetic alignment with the most positive and loving energies that we can imagine participating in. All of life is happening in our own consciousness, and we're aware of as much of it as we desire and allow ourselves to realize. Every form of energy has its unique vibration. With our power of mental and emotional focus, and our freedom to imagine, as well as our ability to feel and know the energetic patterns that we attract, we are the creators of our lives and experiences.

Our creative ability is permanent and present in every moment. We create the vibratory level of our state of being, which manifests as our energetic signature. This is how we believe ourselves to be and to feel. It attracts the same spectrum of energetic patterns that we inhabit mentally and emotionally.

While cruising through time and space, we have developed a lot of personal preferences and attachments to limiting beliefs. These are not part of our reality, because they are based on a sense of lack. In our creative essence, we have only fulfillment of all our desires, unless we choose something else. While learning to be in a high-vibratory sense of awareness in every encounter, we can align with the energy of the heart of our Being in the realm of gratitude, joy and compassion.

Regardless of what may be happening within our range of influence, our energetic alignment creates the qualities of our experiences. We always have a choice of how we think and feel about ourselves and our outer circumstances. If we have no attachment to limitations, we become mentally and emotionally clear, and we can know our inner truth in every moment. Without attachments, our inner knowing is potentially infinite. In each moment the highest potential vibrations for us come into our intuitive awareness.

Having mental and emotional transparency reveals the truth

for us to know in every encounter. By choosing to be clear, we have no constraints upon us, and we are free to engage with the highest vibrations that we desire and can resonate with. Personal fulfillment in every way is possible for us now. As we learn to realize what unconditional love is, we gain the ability to transform negative energies that we encounter into alignment with our positive state of Being, or we can let them dissolve from our presence without our life force.

When we learn to control our state of Being in a perspective of true knowing in alignment with our heart, we become the masters as much as we allow ourselves. Actually we already are masters of creativity, we just need to realize it. It is part of our constant state of being. As we learn to elevate our state of being, we also elevate the state of humanity and guide us all through our ascension to a higher energetic dimension.

Freeing Ourselves from Limitation

Created in the consciousness of the Infinite One, we are fractals of infinite Being in our creative ability; however, in our current human form we have been unable to realize our true essence, unless we have dedicated our lives to inner mastery and have had access to the most enlightened teachings. Even then, it may take many lifetimes of intense practice. Yet, because of the great significance of this time in human history, we have chosen to manifest ourselves now in human form.

As Beings of pure conscious essence of awareness, we could not have the intense experiences that are possible for us as incarnated humans in this density. We would not be able to know fear and suffering, which have given us a deep understanding and feeling of compassion. In our quest for mastery of ourselves, we have been challenged to live in the light and love of our essential Being. We are learning to live in the chaos and turmoil of life on this planet, while developing and maintaining an intentional perspective of enhancing all life in every moment in our thoughts and feelings.

This can be our quest for mastery in this lifetime. By learning to control our focus of attention, we can choose to live in our greatest visions of ecstasy and beauty in every moment. We can choose to communicate with the heart of everyone we encounter and radiate love and joy to all. With the rising resonance of the Earth and our cosmic environment, it is becoming easier to live in love and compassion, even when facing negatively-oriented beings who want our life force.

We do not need to engage with negativity at any time. We can recognize the place of light in negative beings and engage with it. If we don't recognize the light in them, and we need to engage with them, we just need to maintain our connection with positive energy in the confidence that ours is the real world. By holding this perspective, we allow our circumstances to be arranged in harmony with us. As we begin to flow with the creative intention of our Creator, our radiant energy creates miracles around and within us, transforming our lives and our environment.

All of this depends upon our ability to focus the energy of the heart of our Being and hold our focus on the love and freedom of our true Being. Because we have freedom to choose our perspective in life, we have the ability to align with any energetic patterns that we want to experience. By imagining and feeling that we are living in the energetic level of our preferred scenario and realizing that it is real, it becomes real for us. This is an intentional step in conquering our limiting beliefs, and it is possible for all of us. It is our realization that creates our reality, and it is part our greater Being.

Recognizing, Resolving and Transcending Our Limitations

Although we have been subjected to life-diminishing energies for eons, we are on a living planet whose essence is the enhancement of all of life in deepest love and beauty. Parasites do not belong here and are now being brought into the light of public display. They are becoming uncomfortable and incapable of being here much longer. As we become aware that we have acquired limiting beliefs about ourselves, we can decide if we want to keep them. If we choose to transcend them, we can enter a new era of manifesting the creative energy of our true nature.

If we begin the process of resolving our limitations, we can examine ourselves closely to find an obvious limitation, such as, "I'm ashamed of something I did." That is a negatively-polarized state of being. We could think of it as a decision that I made because I had a stronger feeling of wanting what I did. That feeling was aligning with a negative vibration and was an experiment in feeling those energies. They resonated with the energies of the others involved. There was no blame, just an experiment to find out how different energies feel. We have that freedom innately. What is important here is our motivation. Were we enhancing life in every way, or were we intentionally diminishing someone in any way? By judging and mistreating someone or something, we are radiating those energies into the quantum field, attracting those qualities into manifestation in our own lives.

In this way we are the creators of the qualities of our experiences. We are all on a path to somewhere as part of humanity, and we have many choices of ways to arrive there. All we need to know is how we feel in the moment, and we can have intentional control over that. We also have awareness of what we know intuitively, apart from the mind of ego-consciousness. We do not need to think. We just need to realize and know. This is a giant change for us. It means moving from our mind to our heart and realizing the qualities of our essential being.

In the consciousness of our heart, there are no limitations. There is only great vitality and unconditional love, enhancing our awareness. Everything resolves into the creative energy of unconditional love. We do not need to think, because we already know. We can live in the realm of knowing without limitation. It is the energy of the heart of our Being and is who we are. Once we realize that we are naturally and abundantly cared-for, we can express our true interests and passions. There are no requirements upon us, and we can live in a world that supports everything that leads to love and joy. Our experience of this world depends upon the vibratory level of our predominant state of being. This is our major limitation, and is subject to our intentional direction in alignment with our heart.

Examining Our Ascension Process

Because of our incarnation in individual bodies, together with our training to believe that we are separate beings apart from our Creator, we have been unable to know our true identity. We have believed that we are limited to our physical abilities, and that our consciousness is an aspect of our brain. Recently scientists have found that our heart is neurologically much more powerful than our brain. But we do not think with our heart. This is where we innately know what we know without belief or proof beyond ourselves. It is the seat of our intuition and the conveyance of our life force from the consciousness of our Creator. Through the energy of our heart, we have access to greater guidance in understanding and realizing our true nature.

For eons our planet has been enveloped by negative energy, making it a very difficult place to live. Without our life force, the negative could not exist, because it receives no life force from the consciousness of the Creator. When we are negatively polarized, we shut ourselves off from the energy of our heart, and this is the situation of our ego-consciousness. Our thinking mind cannot know what our heart knows, and as long as we are wrapped up in our mental processes, we are missing awareness of our intuitive knowing.

To open ourselves to our intuition, we can transcend the

mind of our ego-consciousness by many processes. We could engage in one of the traditional methods of awakening, such as Tibetan or Zen meditation or serious yoga. We can learn to direct our breath consciously for specific purposes. Spending time in nature and absorbing the vibrations of the Earth is helpful, as is listening to inspiring music and singing. With our strong intention and willingness to accept higher guidance, we can learn to release our anxieties and fears, our limiting beliefs about ourselves, and we can open our sensitivity and awareness to our innate knowing and feeling.

This is a never-ending process of personal expansion. Self-Realization of our true Being is a great step along the way to complete realization of the consciousness of the Creator, of Whom we are fractals of infinite Being. We are free to transcend the chaos and suffering in our lives by elevating our perspective by imagining and feeling ourselves living in scenarios of the greatest love and joy, compassion and fulfillment, until we can realize its reality. We are the only limiting factor that keeps us from realizing our truth. With intuitive realization, we are the energetic creators of the most magnificent experiences of our lives. This level of positive resonance is beyond the reach of negativity and is a state of Being that we can all achieve, because it is who we are. We just need to release our attachment to limitations, and we have to accomplish this absolutely through our intuitive realization.

Transcending Ego-Consciousness

We can be aware of our infinite Being by aligning with the light in the heart of everyone. It manifests as a photonic emission of radiance, and it is the spark of unconditional love that we can innately know and feel and realize. We all arise within the consciousness of the Creator and are imbued with Creator energy through our heart. This is our source of inspiration, constantly

Chapter 3. Resolving the Limitations of Our Consciousness

guiding us to align with its vibratory level, drawing us into greater compassion, love and joy. Our realization of this begins with gratitude. We are here to deeply enjoy living, and we have the ability to do this, regardless of what may be happening in the world around us, for which we can have deep compassion.

Through gratitude for our greater Being, we are given the ability to open our awareness to infinite Being. We can learn to be our eternal presence of awareness in every moment, always aligned with our higher guidance. Universal consciousness arranges our experiences to resonate with our state of being. By opening our realization of this, we can release the distractions of ego-consciousness. The events around us have no consequence for us. All that is important is how we personally use our attention to guide our vibratory level. While being mindful of the quality of our personal vibrations, we can open ourselves to our higher guidance.

Flowing into us is the creative life force that abundantly provides for us in every way, if we allow it. Our attention in gratitude is required for this awareness. We can choose to shift our attention into a clear presence of awareness. Our personal dramas and needs can transform into personal fulfillment, once we transcend our ego-consciousness of doubt and fear. The quality of our personal reality depends upon our attitude and perspective, as well as our mental and emotional actions and reactions. These are the creative powers designing our experiences.

Our polarity and vibratory level are our creative energies. As often as we focus and align with the vibratory expressions of any energetic pattern, we create that level of experience for ourselves, regardless of what may be happening around us. Being aware of how we feel in every situation, we can recognize if we are in ego-consciousness or beyond it. Beyond it is the truly creative realm, because we can use unlimited life force. In ego-consciousness, we have limited ourselves in how much divine energy we can embody, and much of it we have given in alignment with negative, parasitic energy.

When we can realize our expanding Self by being our clear presence of awareness in love and joy, we are in alignment with our infinite Being and are unlimited in our creative ability. We can all have this realization. This is the direction that humanity is being drawn toward, as we gain ascension-consciousness into a higher energetic dimension.

Becoming Transparent in Our Radiance

When we are able to release our attachments to limiting beliefs about ourselves, we become transparent in our relationships and how we think and feel about ourselves. The life force that we receive from the consciousness of the Creator glows in our personal radiance, because we do not restrict it or give it to negative entities, who are parasites feeding off of our etheric essence. They feed off of our fear and depend on our limiting beliefs to trick us into energetic alignment with them. Attempting to hold us in fear, their propaganda is everywhere.

We have the ability to transcend all negative energies through our focused alignment with joy, compassion, gratitude and appreciation for our infinite essence of Being and unlimited creative powers. We can develop sensitivity to the energy of the heart of our Being through our intuition. Until we release our limiting attachments, we do not realize our true essence. To our ego-consciousness, this release requires a leap into the unknown, because the ego does not know higher guidance or understand how miracles are created. Yet we can transcend ego-consciousness through intentional introspection into our attachments to limitation.

By examining our limitations in the light of what we know intuitively, their true nature becomes apparent. They are all based in fear and are negatively polarized energetically. The greatest fear is the expectation of death as physical and conscious termination. Quantum physics has proved that conscious termination is impossible, because we participate in universal consciousness beyond time and space, along with all conscious beings composing everything we know, down to the subatomic entities that we recognize as photons, electrons, protons and neutrons. While our bodies' constituent entities maintain their identity and consciousness, our physical forms can change and become other forms.

Consciousness is constantly expanding. It never disappears or is diminished. Because we live in universal consciousness, we have access to as much of it as we allow ourselves to be aware of. Only our self-imposed limitations keep us from a life of great awareness and creative ability. Once we resolve our attachments to limitation, we gain true freedom and can live beyond the current dramas of humanity, while still participating in our chosen way of life. As we begin to realize the power of our divine life force, our lives become miraculous, and we are able to live in the radiance of unconditional love and compassion for all who are still limiting themselves.

The Next Step in Expanding Our Conscious Awareness

In the journey toward personal fulfillment, expanding to include all of humanity, we have much training and programming of personal limitations to overturn and resolve. Our next stage of evolution is the expansion of our consciousness to the realization of our unlimited choices in expressing our creative ability. This depends upon our mental and emotional shift in polarity and vibratory level in alignment with our intuitive knowing, the expression of the energy of the heart of our Being.

Because we have been trained and persuaded to accept entrancement within a limited scope of awareness, and it is the turning of the ages into a time of greater knowing and awareness, we are being drawn by the rising energies of our cosmic environment to open our awareness to our greater Being. In order to make this shift within ourselves, we can drop our attraction to negativity and begin to transcend the realm of duality into a dimension beyond polarity and fear, a level of consciousness expressing unconditional love and joyful living. Here our choices can expand to an infinite number of ways to expand and enhance life everywhere we focus our attention.

We still have the choice to delve into the negative, but we have no attraction for the realm of belief in fear and mortality. We've been exposed to that entire realm and can begin to remember our true Self, unlimited by space/time in the realm of duality. We can be grateful for the experience, because this kind of knowing would be impossible in our infinite consciousness. We could not hold these limiting beliefs about ourselves. They are unbelievable for our greater Being. And so we created a realm of limited consciousness in order to experience negativity. Now we know all about it personally and can have deep compassion for anyone stuck in this dimension, as if it is real.

Because everything is an expression of electromagnetic energy in infinite variations of swirling patterns, from the sub-Planck subatomic to entire universes and beyond, we can participate in any of it by our recognition and realization. We are the creators of our own experiential qualities by the focus and alignment of our attention with the wave patterns that we recognize in our perception and imagination. They are both closely related. The only difference is our realization of the reality of one, but not the other. The other, imagination, is not believable as real in our ego-consciousness. In order to change this, we can approach it from a perspective of openness and clarity of mind and emotions, closely examining the feelings of the energetics that we're dealing with. We know when there's a tinge of fear or doubt, and this is when we can align our focus with our intuitive knowing, and feeling the energies of our heart, which provide the guidance we need in any situation. By aligning with its quality of energy, we are guided into the dimension of expanding awareness and greater realization of our creativity.

Elevating and Inspiring Our Lives

Our needs are a result of limiting beliefs that we need food, money and shelter. Our health and well-being are a matter of our energetic vibrations, just like everything else. Our bodies are made of consciousness. Every bit of them. Each atomic and subatomic entity is a conscious being, organized by our personal essence into our physical expression. Our consciousness is capable of expressing whatever we direct in alignment with the quality of our personal vibrations.

Our limiting beliefs keep our consciousness from expressing its potentially unlimited manifestations. We do not need material. We only need to know deeply that we are abundantly cared for. All we need to do is to realize it. We make it real in our experience through our realization. This is an example of what quantum physics has shown to be the way the empirical world comes into existence for us. We empower it with our life force through our nearly-constant realization.

As we begin to realize our unlimited essence, we can drop any personal need that we may have believed in, because it has already been fulfilled. We just have to realize it, when we align our vibrations in polarity and vibratory resonance with the energetic expression that we pay attention to. Because of the energetic density of the empirical world, and the way we sense it, we are challenged by powerful distractions to pay attention and maintain an alignment with our intuition. When we can do this, our lives become wonderful in every way. We vibrate in resonance with the consciously-creative life force that flows through us and connects us with universal consciousness.

When we can be deeply passionate about something, we create powerfully attractive energetic patterns that can resolve limiting beliefs, if this is what we want. By envisioning and feeling the fulfillment of our lives as we imagine, we create the energy patterns that enable us to realize its reality. We are the creators of all of our experiences consciously and subconsciously. By being passionate visionaries, we can change our subconscious programming and resolve our limiting beliefs. Everywhere we hold onto fear and doubt, we can choose to transform our awareness to love and compassion. Nothing else has to change, only our awareness. What we realize is what we experience.

Once we recognize what negative expressions are in their energetic essence, we can resolve them, with courage if necessary. We can train ourselves to be elegant, compassionate and understanding in every situation. These are characteristics of a more expansive way of living and are in alignment with inner guidance. By practicing being grateful, compassionate and loving in the moment, we eventually realize that we're living in an energetic dimension beyond negativity.

4.

The Importance of Our Intentions

Creating Our Life Experiences Intentionally

If we can affirm that we are expanding our awareness in alignment with our true heart, we are moving beyond our limitations. If we are resolving our limitations to Self-Awareness, our imagination can grow in brightness and joy. We can know how to be positive always, guided by our intuition. Once we have confidence in our intuitive knowing, we can resolve our ego-consciousness. As we are connected to infinite awareness and knowing through our intuition, we no longer need ego control of our awareness.

We created our ego-awareness by believing in the reality of our limitations without higher guidance. This is a human fabrication and is supported solely by human life force through our beliefs in its reality. It is attached as a density in our consciousness and has no existence beyond what we give it through our focus on its reality.

For most of humanity, ego-consciousness is all that is recognized as our identity. There is little interest in looking beyond time and space, but that is where our true Being resides. We can recognize ourselves as eternal present awareness. We have always been our individual conscious awareness, and we will always be. Time and space are irrelevant, except within the compartment of traditional human consciousness. We can participate in this level of awareness, but we do not need to be limited to it.

By intentionally being positive with the guidance of our intuitive knowing in absolute confidence, we can live in perfect peace. In this state of being, we can master our creative power through knowing and feeling what is life-enhancing for all in every moment. We can recognize the eternal light in each other's eyes and in all conscious beings.

Without ego-consciousness, we become limitless and at the same time able to be present as our ego-selves, but with a higher, more penetrating perspective. We can realize that the spectrum of traditional human energy in the realm of duality is a limited compartment of our consciousness. It requires our alignment with its resonant frequencies and polarity.

Once we become completely positive in every way, we find that we really do have the ability to create everything we need and that our heart desires. Since there is no fear or doubt, all of our creations manifest in our experiences at the vibratory level of their creation in our thoughts and feelings. Our alignment with the life-enhancing vibrations of our heart energy draws only positive experiences into our lives, which we can continue to be grateful for.

Living in Duality from a Perspective in Eternity

There are many steps that all have to work together to make the change in perspective from duality to only positive. Being pure of heart is a necessary step on the inner path. Being open and non-judgmental is another. Wanting and intending to be kind and understanding always is another step, and forgiving ourselves of everything. Our life was partially just experiences in negativity for us to balance with the loving and life-enhancing state of our natural Being. In our true Being, we are far beyond ego-consciousness in expansive awareness, and we can transcend the ego's limitations by attuning to our intuitive knowing.

In every scenario, we can direct ourselves to imagine the most life-enhancing energy that may be present. By practicing this, we become aware of higher vibrations all around us. Eventually we can transcend even body consciousness and just be pure Self-Awareness. As we gain sensitivity to our inner guidance, we gain confidence in ourselves to be able to hold our focus on positive, high-frequency thoughts and feelings from an objective perspective in every situation.

Intuitively we receive guidance in ways that our higher Self and our guides and angels feel that we can understand. This guidance is accompanied by events and symbols in our lives and in our dreams. Always present immediately in every moment, it is our deepest knowing, and there is never a pause for wonder or doubt. We have the ability to transcend our past negative creations with our intuitive guidance

Training ourselves to be present in awareness, open and clear of encumbrances helps in developing intuitive sensitivity. If we face crisis situations that demand our attention, we can still choose to focus on our inner guidance and immediately know how to interact in the moment. In this way we radiate compassion and gratitude always and continue to create goodness and beauty in our energy field all around us. If we develop a clear connection with our intuition and have a strong intention with absolute confidence, we can create whatever our heart desires. It is our energetic expression in greatest clarity.

The Depth of Our Consciousness

Mastery of ourselves requires absolute alignment with pure present awareness and intuitive knowing. This is possible for all of us, provided we are strongly motivated and open to higher energetic expressions. We have a compartment of duality in our consciousness that is bound by our limiting beliefs about ourselves. If we can resolve these through awareness in alignment with compassion and life enhancement, we can transcend the compartment of duality into unlimited Being in eternal present awareness, able to be anyone anywhere in eternity. Everything is our choice. We have no limits to our personal awareness and creative ability.

Because we are fractals of the One who contains all consciousness, we are part of unlimited consciousness. Understanding this leap in awareness from duality to only positive on a high level, is possible. Any hint of doubt or fear is negative and is part of the realm of duality. Whenever they arise, we can transform them with confidence in alignment with our intuition, which is an expression of the heart of our Being. Any deep urges that come from within must be faced and resolved intuitively. In our essence, we are pure divine Being. Intuitively we know who we are, and it is our intuitive knowing that guides us into our eternal present-awareness in unlimited Being.

We have been projecting ourselves as human persons playing our destined roles for this experience and were given a largely hidden subconscious innate part of our consciousness, so that our bodies can function without too much life-diminishing energies from our ego-conscious self, who has little awareness of higher guidance and is attracted to negative scenarios. The innate part of our conscious self has only deductive reasoning ability. It cannot think inductively, as our ego-conscious self can. It's designed to function perfectly in every cell, but it can be deeply influenced by our traumas and deep fears. These cause malfunctioning in the innate consciousness, leading to physical defects, pain and ultimately physical death.

If we decide to take a different path from duality, it can take us ultimately to our unlimited Self Awareness. It requires a strong intention to know who we really are and an openness to our inner knowing. If we can train ourselves to be positive, loving and compassionate in every moment as a state of being, we can create that energetic aura, which attracts resonant energetic patterns with similar mental and emotional stimulation. By being absolutely positive in our lives and following our intuitive guidance, we can expand our awareness into our true Being, our unlimited present Self-Awareness.

The Workings of Our Intention

As more of us awaken to the possible reality of living in a realm of joy and beauty, we lose our interest in the realm of duality, withdrawing our life force along with our attention. This is resulting in the collapse of the world that we have been living in energetically. This realm is held in place by the consciousness of humanity, and our alignment with its energy gives it reality for us. For us to jump into the world of love and joy, we need a strong desire and intention. Once we achieve this level of vibration in ourselves for sustained periods, we provide a strong positive vibration for humanity, stimulating others to be interested and attracted to the recognition of a higher energetic dimension, which we can be living in.

We can read the energy in the moment by recognizing if it is completely life-enhancing, or if someone is being diminished by it. In order to master this discernment, we can transcend our ego-consciousness by intentionally paying attention to our intuitive guidance, which is always life-enhancing in unconditional love, and is our connection with universal consciousness. We can intentionally align our thoughts and emotions with this positive-feeling vibratory level, even when we're facing strong negativity. Our state of being is always our choice, and our choice controls the quality of our experience. Circumstances arrange themselves to accommodate our vibrations.

If we are completely positive and confident in enhancing all of life in our every breath, we create the fulfillment of all the desires of our heart. There is nothing limiting or destroying our creative thoughts and feelings, because we no longer have doubt or fear. This can be understood purely from recognizing the energetics involved, the polarity and vibratory resonance. All of our intuitive guidance comes with a positive polarity and high-frequency resonance. We naturally feel good about it, and it is good for everyone involved.

With a strong intention to be our true Self, we can attract all the energy we need to make the leap in consciousness to be only positive and to realize the life of true beauty, abundance and freedom that is available to all who can recognize its reality.

Our Power of Free Will

On the path to expansion of our awareness, we can learn the real value of having free choice and how to use it to attain Self-Realization. Free choice is most important in feeling our emotions and knowing our thoughts. Thoughts and emotions have a vibratory level and polarity. In our world of duality, we are able to choose our thoughts and emotions and experience their vibrations. Our current state of being is a result of our use of free will. It is what we accept as well as what we create. It has to do with our energetic alignment, beginning with the polarity we live in. If we expect to have our consciousness diminished and terminated in death, we live in fear. This is the situation within the consciousness of humanity.

Since we have been thoroughly trained to stay within the limits of the consciousness compartment of humanity, choosing to be only positive always is a difficult leap in consciousness.

The grip of fear on our psyche is deep and strong. Our ego-consciousness can comprehend only personal separation in being and limited conscious awareness. Higher guidance is beyond it, and positive polarity is a death sentence in the face of grave threats in the experience that the ego recognizes as real.

When we are aligned with negative polarity of any kind, we cannot believe in our infinite Self. That part of ourselves is not real for us. To recognize our true Being, we must open our awareness and be receptive. The opposite polarity from fear is love and joy. When we are able to be positive always, there is a natural fulfillment in every area of life. Without doubt or fear, we have only confidence in everything we imagine that our heart desires.

Our intuitive knowing is always present for our awareness, awaiting our intentional alignment. Once we recognize our inner knowing with confidence, we no longer need our limited ego-self for guidance and reaction in living. We can become transcendent in gratitude, love and compassion in our encounters. We become free to follow our heart's desires.

Every step of the way to awareness of infinite Being requires a choice. To continue expanding, we must continue to choose to be positive, even when our ego shouts danger. In the realm beyond polarity, which we can participate in, only positive energy exists. Negative events cannot happen, because no one creates and aligns with them. Even while we live in the world of duality, we can choose to be only positive always and thus transcend negative events in our lives.

Leaving the Matrix

For eons humanity has been controlled by parasites. They could not have power over us without our permission, which they have taken through dissimulation. We can transcend our experiences with them by withdrawing our permission and our life force. Sovereignty is awaiting our conscious recognition and is part of being only positive.

When we are positive, we cannot have fear, which is negative. We cannot have significant contact with negative energies, because they receive nothing from us. They need our fear to exist, in the sense that we subject ourselves to them out of fear. Since we are then aligned with fear, we are giving our life force to the existence of negative energy.

The parasites have kept us in a hypnotic trance, not knowing that we can leave the negative realm whenever we become positive throughout our Being. It is an intentional choice to transcend our ego-consciousness and to open ourselves to expanding awareness of the energy we receive in the consciousness of the Creator. This is what's natural to us. It is life-enhancing in every way and is what we can know intuitively. At this level of consciousness, there is no alignment with parasites, and no interaction.

Our eternal Self-Awareness is being aware beyond time and space, which is only positive and is the source of our human character. Once we clear ourselves of our attachments to our limiting beliefs about ourselves, we become emotionally and mentally clear. Our expanding consciousness can open us to realizing our innate abilities. In our higher-conscious awareness we are unlimited in every way. We are our present awareness with infinite abilities, and we have chosen to express ourselves as our human being.

These realizations can transform our lives into the experiences that we intentionally choose. Because we can be aware of the quality of the energy at the heart of our Being, we can express our life-enhancing energies in our experiences. We can have deep understanding in every circumstance that we encounter, and we can participate in great awareness within the One universal consciousness.

Continuing to Expand Awareness

By the energy that we are radiating into the quantum field, we are expanding consciousness. Through unfocusing our attention and just spacing out in meditation, we can open ourselves to expansion by asking for it and aligning with its level of vibration, which is what we can feel emotionally. We can imagine wonderful scenarios, feeling joyful, as much as our limitations will allow. As we continue to search for purer feelings of love, we raise our personal energetic resonance, and we can realize that our awareness is expanding beyond the empirical world.

When we can be emotionally and mentally clear, we can learn to call up emotions as we desire to feel them. We can also be aware of what we feel around us. By paying attention to our emotional feedback, we can learn to read the energetic quality of everyone and everything that we encounter. This is part of our intuition. When we are aligned with the enhancement of all life, and we are looking and feeling for high-vibratory experiences, they come into our awareness.

All patterns of energy already exist in the quantum field. For us to experience them, we can open ourselves by asking for experiences and imagining that we are living in them. When we can believe that they are real, they come into our experience. This is our creative ability. It is the reason why it is important to resolve the beliefs in our limitations that we hold. When we hold doubt and fear and other negative vibrations, we create experiences that stimulate those feelings in us.

If we want to have only positive experiences, we can be mentally and emotionally positive, regardless of the energy that we face. Our lives are constantly rearranged by the qualities of our vibrations, and our circumstances change accordingly. If we stay within our limitations, our lives may not change much, and we won't evolve spiritually. This seems to be comfortable for many.

For true expansion of our conscious awareness, we can focus on infinite Being beyond time and space. This may require some practice to imagine as a present state of being, but as we are able to align with our intuitive knowing, we can open ourselves to our eternal present awareness.

Transforming Ourselves in Alignment with Our True Being

Quantum physicists have shown that there is a universal consciousness, and that it is the creative source of everyone and everything that exists. It is inferred that this is the consciousness of the Being that we may call the Creator. Everything exists within the consciousness of the Creator, including us. It is further inferred that our consciousness is the consciousness of the Creator. Logically, there is no other possibility. If the consciousness of the Creator creates and envelops everything, there is nothing outside of it. We cannot exist apart from Creator consciousness. This means that our awareness dwells within Creator consciousness.

Because we exist within universal consciousness, we share in the entirety of it in our potential awareness and creative ability. We are fractals of the Creator, identical in every way, flowing seamlessly and mathematically perfectly throughout the cosmos, like an MC Escher drawing. Because we each have the ability to create and destroy by virtue of our free will within the realm of duality, we have agreed to operate subject to our own doubts about our identity. We keep ourselves from realizing our true creative and destructive ability. If we decide to become life-diminishing, we close ourselves off to our conscious life force and must become parasitic on others. We see this throughout our society.

Because of the powerful influx of light enveloping the Earth at this time, and the rising, life-enhancing vibratory resonance of Gaia, parasites are becoming desperate and even insane. They need to feed on our life force, which they do through eliciting energies in us that are compatible with their own in fear and negative polarity. We can make ourselves immune to them by being only on the vibratory level of kindness and compassion.

In our true nature, we are only positive in unconditional love and life-enhancing intentions. These are qualities of the universal Creator consciousness. As we train ourselves to be always only positive, we begin to trust ourselves to be true to the life-enhancing creative energy of universal consciousness. Without a negative orientation based in doubt and fear, we no longer cancel out our positive creative intentions, and we open ourselves to experiencing joy, freedom and abundance. We no longer need to participate in the chaos and suffering that appear to be occurring around us. In our own lives, we can experience the most powerful creative force of the unconditional love of universal Creator consciousness.

Living in Divine Service

Many of us are embodied on this planet at this time to express our light, love and compassion to all of humanity and to Gaia, our Earth Mother. To transform our energetic environment into a true expression of the unconditional love of the divine Creator, we can work in alignment with the forces of nature and the intergalactic community. We can do this with our personal perspective and way of being in our interactions with others and in the energy that we hold and radiate in our energetic signature.

This is the most important time in the history of the Earth and of humanity. We are passing through a turning of the ages, which occurs at the beginning of every 25,920-year precessional rotation of our planet, which occurred exactly on December 21, 2012, when the Earth conjuncted our Sun and the center of our galaxy. We are entering a time of ascension into a higher dimension of living, a dimension of positivity and high vibrations. The negative, dark energy that has enveloped us for eons has been diminishing ever since, and now it is losing its last bit of power over those who still accede to its propaganda and dictates. There are still appearances that can stimulate moments of fear in us, but they are without substance.

Any appearance and feeling of negativity can be overcome by keeping our awareness on our intuitive knowing. We can depend absolutely upon the life stream coming to us intuitively in every moment from the Source of our Being. We are never alone in the essence of our consciousness. Our connection with the divine is always actively expressing everything we need to know to live in gratitude and joy, and we have the freedom to pay attention to it or not.

Everything we need is within our own Being, but our limiting beliefs have kept us from being aware of our inner knowing, which comes to us in ways that are unique to each of us. To be aware of it requires our intention and receptivity beyond our ego consciousness. Because we are the essence of our Creator, we must ask within ourselves and then be receptive to the first impressions that come into our awareness. We can practice to be certain of our inner guidance. It is not a mystery. Once we resolve our attachments to our limiting beliefs about ourselves, we can be open to what we truly know.

Resolving attachments to our personal limitations, and being in alignment with our intuition is the solution to higher awareness that enables us to create the enhancement of our lives and the vitality of everyone and everything around us. Each of us has the potential of unlimited awareness and creative ability in loving service to the Source of all that exists.

Becoming Open, Clear and Receptive to Inner Knowing

All around us there are great forces designed to attract and compel us to assume a perspective that will either enslave our spirit or grant us ultimate freedom. We are in the transition between energetic eras in human history, and the changes that are coming about are beyond our historical imagining. We are being required to choose either to remain in duality or to transcend our limitations and open ourselves to living in the consciousness of great love and joy.

If we choose to remain in duality, we will continue to experience the entire spectrum of human life that we have enjoyed and endured for eons, with all of its pleasures and pains. This will continue on another planet that will seem to be just like the Earth has been, but the Earth is changing. All who remain here will live on a regenerated Earth of beauty and abundance, with advanced technology and inter-galactic connections. The energy will not support predators, only heart-felt goodness and creativity. To live in this environment, we must be able to imagine participating in it and believe that it is real.

We are the creators of our experiences, and we have been limiting ourselves to the realm of duality, believing that it is real, and we have closed our imagination and feelings off from our greater awareness. Once we realize that our ruling elite are duplicitous and parasitic, we can begin to recognize how they have used our own creativity against us for their benefit. It is done through stimulating fear in us. Our fear aligns us with their negative vibrations and allows them to take our life force, which they then use to control us through threats, intimidation and limited monetary access.

By being irritated and angry toward them, we continue to align with their negative energy, allowing them to continue to create fear in us, even on a limited level. Once we decide to be only positive, we no longer give them our resistance or support. We can be grateful for the challenges they have given us to learn to control our state of being. By responding to them with acceptance and forgiveness, we remove ourselves from their energetic realm, as if by magic.

Within each of us is a connection with the essence of all living consciousness and awareness. It is our heart-mind, which we can know as our intuition. In order be aware of it, we must intentionally want to realize our inner knowing in clarity and gratitude. It is our opening into realizing our infinite Self in the unlimited consciousness of our new world. By practicing being sensitive to our inner knowing, we gradually acquire mastery of ourselves and always know everything we need to know to experience the lives we are destined to enjoy in alignment with the rising energy of the new era in human history.

Aligning with the Energy of Our Heart

Sometime we all feel a strong attraction for the deepest love and greatest joy. These are states of being that we can align with energetically and emotionally. Our greatest love is waiting for our resonant alignment, so that we can feel throughout our awareness the vibrations that connect us with universal consciousness. There is only positive energy shared throughout the cosmos by the creative intent of universal consciousness. Our participation is voluntary.

As shown by quantum physics, universal consciousness is real throughout the cosmos. It envelops all conscious beings, enlivening us to the limits of our beliefs about ourselves. As our limitations come into our realization, we can examine them closely and feel their emotional depth. If there is any level of fear, they are creations of our ego-consciousness. If there is only positive energy, there is nothing to be afraid of or threatened by.

This level of consciousness exists in the higher vibrations of human consciousness, and it is available for our resonant alignment. By recognizing the energy of our greater heart, we can be drawn into an awareness of the pure love and joy radiating throughout our being. We have the ability to choose to align with this energy or to disregard it. If we intend to achieve alignment with the high-vibrations of love, we will elevate our perspective about everything and begin to come into realization of our true abilities. We change our entire life, with the radiance of gratitude, love and joy permeating everywhere.

Our lives fill with goodness and compassion, in alignment with our heart-felt intent. In this way we remove ourselves from karma and declining life, because we express the energy of positive creation. We are engaged in the positive enhancement of all conscious beings. We can know that our every encounter is part of the flow of enjoyable experiences. Higher vibratory states of our awareness open up experiences for us that were beyond belief for ego-consciousness.

We can realize that everything is getting brighter. We're moving through a gigantic conscious photon field that is raising the resonant frequency of the Earth, as well as ours. All of the most deeply-hidden negative energy is being brought into the light for examination and resolution or dissolution. We see this energy manifesting throughout humanity as divisiveness, hatred, chaos, conflict and destruction. It is running its course, until humanity no longer supports negative energy with our belief and engagement.

One-by-one we are moving beyond limitation, as our inner light grows brighter and we become more positive in every way. We can learn to pay attention to the feelings of our heart in every moment. This is life-enhancing energy that feels good. As we intend to come into resonant alignment with our heart-felt energy, we can enhance our awareness in the beauty and vibrancy of nature. We have the ability to achieve this level of living whenever we choose it.

Living in Creative Power

Although we have been tricked into enslaving ourselves to a group of psychopaths, we do not need to continue to be victims of our circumstances. Our circumstances reflect our own perspectives and how we feel about ourselves. As long as we are in anger, doubt and fear about our situations and abilities, we are creating circumstances that manifest those feelings and thoughts. Our experiences are caused entirely by our own energetics, because we are fractals of universal consciousness and are inherently creative in everything that we think and feel. The only time we're not being creative is when we're just being present in awareness.

If we feel that others are diminishing our lives, there is only one way to transform our situation, and that is by changing ourselves with our own perspective and attitude. Our life situation always begins with our free will choices. We decide what we want to focus on and how we want to feel about it. We have the choice of acting with creative intent or reacting subconsciously as a result of misunderstanding the nature of our essence. We have been taught to react to those who diminish us with anger and even the force of weapons. This aligns us with the negative vibrations that we accepted, perhaps inadvertently, that brought about the situation that we are reacting against and only strengthens it with our own life force.

In order to resolve the situation in our own experience, we can align ourselves with the energy of our intuitive knowing. This is what we feel and know in the heart of our Being. It is the source of our consciousness and our life, the creative energy of our true essence. It is our conscious connection with the universal conscious awareness of our Creator. Although we have closed ourselves off from this awareness in order to experience the duality of living as humans, we can transcend our ego-consciousness when we have a strong intention to expand our awareness.

If we want to live in the truth of our Being, we can choose to resolve our limiting beliefs about ourselves and to open our awareness to inner knowing of our unlimited abilities. These flow to us when we align our mental and emotional vibrations with the unconditional love and joy of our deepest creative essence. We are designed to be unlimited in every way, to be infinitely creative. In our deepest consciousness we can realize our true stature with confidence and gratitude in order to transform our lives into experiences of miracles and greatest love in eternal, infinite awareness.

Realizing Our Higher Destiny

As our natural environment becomes more positive with higher vibrations, we are being drawn into greater love and fulfillment. We are being invited to withdraw our attention and alignment from the psychopaths and their destructive energies, as well as our own deficiencies, and pay attention to the energies of our heart. When we find ourselves locked into dire situations, we can escape by closing ourselves off and beginning to laugh deeply and loudly. The more we laugh, the more we feel like laughing. We cannot be negative at the same time we are laughing whole-heartedly. Laughter carries us into positivity, invites us to breathe deeply and elevates our mood. If we do this with others, the energy gathers strength. Dancing and singing heart-felt songs can have the same effect. They free us from the clutches of negative energy.

Once we are able to be in a positive space, we can dream of love and beauty and the fulfillment of our deepest desires. We can feel ourselves enjoying everything we can imagine in gratitude and confidence. If doubt and fear creep in, we destroy our visions and sabotage our intentions, and we need to change our perspective and start over. Without confidence and gratitude for the fulfillment of our desires, our visions and intentions become just fantasy without substance. Every bit of energy in our thoughts and emotions is significant in creating the experiences of our lives.

With the heightened energies swirling around us in the natural world, great changes are being brought about in all aspects of our lives, and we are in the position of constantly choosing, consciously or subconsciously, the energies we want to align with. We are in the process of creating our new world by the energetic choices we make every day. This is our opportunity to expand our awareness into greater consciousness.

We all have the intuitive guidance of our heart, which draws us into the most fulfilling energetic patterns that we can experience. If we have the desire to be aware of our intuition and the energy of our heart, we can use our natural creative abilities to direct the quantum field to manifest the life that we truly want and the greatest awareness that we can open ourselves to. As we all learn to do this, we prepare for the great ascension of all conscious beings into unconditional love and universal consciousness.

Realizing Our Creative Potential

We can ask for awareness of being in the consciousness of the Creator, the infinite One that is All, the Source of our conscious life force. By aligning in resonance with life-enhancing positive vibrations, we can realize the energy coming through our heart. Our intuitive knowing opens to us, and we can have access to universal consciousness. We can create whatever we want by realizing it and feeling ourselves interacting with it. When we do this on a heart-felt level, and we recognize its reality, it becomes real in our experience.

We are the consciousness of the Creator, as much as we allow ourselves to be. By being open to recognizing our shortcomings, we can realize how we keep ourselves from receiving the unconditional love that every encounter can bring us. If we harden ourselves against experiencing negative energies, we actually resonate with their vibrations and hold them in our experience. The way out of this enslavement is to raise our vibrations in the positivity of gratitude, joy and compassion and to intentionally transform the energy that we have been holding.

We can control only what happens in our own consciousness, and this we can control completely. We are designed to create whatever we can realize as real in our feelings and beliefs, actual or imaginary. Our energy signature radiates our polarity and vibratory level into the quantum field for manifestation into our experience. By staying in the energy of our heart, we can maintain a confident perspective of love and compassion, regardless of outer circumstances. By being confidently positive in ourselves and in our encounters, we can transcend our beliefs in our personal limitations.

In the consciousness of the Creator, we are unlimited, and we can realize this by aligning with the energy of our heart and our intuitive knowing. This awareness is beyond ego-consciousness and requires transcending our limitations. It is an intentional reorientation of our perspective and can transform every aspect of our lives based in negativity, which does not exist in higher consciousness. We can live from one miracle to the next in confidence and gratitude. Our natural state of Being is abundance, freedom and joy, as much as we allow for ourselves, for we are the creators of the quality of our life.

Working with Our Personal Vibratory Level

With infinite life force flowing throughout universal consciousness, its direction is toward creating and enhancing all conscious beings. This is the energy we are being attracted to, and we are uncovering some of our previously-hidden abilities in transforming energetic patterns. If we can recognize them, we can also imagine them from a perspective of love and compassion. When we recognize the positive energies of love always present, the vibratory level rises in everything in our presence. Lower vibrations either come into resonance or become unstable and dissipate; however, other beings can continue to sustain negativity with their own life force. When this happens, if we remain positive, life rearranges our circumstances, so that we don't have to encounter negativity.

If we can master aligning ourselves with the energies of our heart, we can achieve constant present awareness. This releases us from the constant stress of low-level fear and empowers us with greater life force and radiance. When we expand our awareness into the realm of joy and gratitude, we can live in the moment, knowing we are always cared-for abundantly and doing what we are intuitively prompted to do.

The personal energy signature of most humans does not change much during a lifetime, but when we begin to raise our vibrations intentionally, we also change our energy signature. This elevates the quality of our life experiences and makes it easier for us to be present and grateful. Once we choose to know our true essence and hidden abilities, we can begin to resolve the self-limiting beliefs that we have been programmed with. Upon close inspection, we can recognize and feel their unbelievability, leaving us unlimited in our awareness. This is a process we can work through successfully.

Realizing that we have the ability to choose our polarity and vibrations in any moment gives us the power to elevate our energy signature, which opens up a world of positive experiences. Learning to control the focus of our attention and to change it at will, along with being able to call forth emotions and change them at will, eventually gives us mastery of our lives. When we can do this while following our intuition, we can direct our creative power to the enhancement of all of life.

Understanding Thoughts as Prayers

Our every thought is a prayer, because we are the creators of the qualities of our lives. It is not the passing thoughts that we observe without judgment, but the ones that we connect with in some way, either by resistance or attraction. Although we live within a limited bandwidth of frequencies, we know from quantum physics experiments that every sub-atomic wave/particle in our bodies participates in universal consciousness. We haven't been able to access this expanded awareness, although we're potentially capable of it, because we have trained ourselves through our families, cultures and businesses to limit our awareness to the energetic patterns of duality and empiricism.

We live within the realm of good and evil and what our senses perceive. This entire world of human experience is changing. The sun is getting brighter, and the positive polarity of the light is strengthening. The polarity of the Earth is shifting to only positive at the center and her resonant frequencies are rising. All of the darkest negative energetic patterns are coming into the light for transformation or dissolution. It may appear as a dark time for us, as all of humanity detoxifies, but the direction of our life force is toward the light and positive energies in the vibratory level of love and compassion.

It is no longer experientially necessary for us to participate in duality. We keep it in our reality with our negative thought-prayers. When we judge ourselves and others negatively, when we feel stress, doubt and fear, and when we believe that we are victims, we are focusing our attention on negative energies. Because we are created to be energy modulators in our consciousness, we radiate this negative energy into the quantum field for empirical manifestation in our experience. We attract the same kind of energy that we pay attention to. This is why advertisers want our attention.

Prayers that are only words without attention are like random thoughts that pass through us and have no effect, but thoughts that are intentional are creative. When we focus on what we don't have, we create not having it. When we focus on enjoying what we want, we can be grateful and begin to believe in its presence. By focusing on living with the kind of energy that we want, we attract it into our lives. By focusing on being loving and kind, we attract this kind of energy.

Whatever level of polarity and frequency we vibrate at determines the quality of our experience. If we vibrate at the level of gratitude and joy, we can see the unconditional love of the Creator in even the dark beings. This perspective enables us to inhabit a dimension of energy that is beyond the vibratory alignment with negativity, and the dark force cannot penetrate it and does not exist here. We can choose to be only kind and loving in all encounters and confident in all aspects of our lives.

It's time to play and enjoy life. When we choose to focus on experiencing the unconditional love and life-enhancing energy of the greater Being that we are part of, our thought-prayers can create wonderfully magical experiences for us and elevate us beyond the experience of negativity.

Realizing Higher Consciousness

Connecting with the energy of our heart and aligning within intuitive knowing is the most important process we can engage with. We can be completely Self-directed, living in gratitude, love, abundance and freedom. As long as we can maintain a positive perspective and feel ourselves in the high vibrations of joy and compassion, we enable ourselves to live in a higher dimension of light, that provides mastery of the human situation.

Many challenges await us, but we have had sufficient experience to understand what is happening. No longer can the negative elite control us, if we do not psychically feed them our life force through our fear, anger and frustration, which they do their utmost to engender in us.

If we seek within our own knowing, we can begin to realize our eternal Being, our present awareness beyond time and space. Our experience of this knowing opens up to us as a result of our intention. If we ask for it, we have much help available in the inner realm. Each of us has guides, angels and ascended Beings, as well as our own higher Self, who can open our awareness when we are ready for realization of our eternal presence.

What is required of us is being quiet in an environment that is inspiring, turning off all outer stimulation, except possibly for quiet, uplifting music or the sounds of nature. We can begin with a few deep, rhythmic breaths. At first, we can just be aware of the thoughts and emotions that pass through us, as we turn down the stress, anxiety and fixations of our lives. We can practice observing without engagement or involvement. We can also learn to be intentionally creative, imagining and feeling ourselves living in higher-dimensional scenarios. Ultimately, we can move into alignment with the consciousness of the Creator in unconditional love and joy. The more we can do this, the easier it is to release our attachments to personal limitations and to become intentional in every moment.

Our goal can be mastery of life in a heart-centered environment, participating in a positive, high-frequency state of Being. This spectrum of consciousness exists for us as soon as we can recognize it and realize that it is real. This is our great challenge in learning how to expand beyond our limiting beliefs about ourselves. If we can keep intending to align with the life-enhancing energy of our heart until it becomes habitual for us, we receive inner knowing of how to do this. Everything we need is brought to our attention. Mostly we've been ignoring this guidance, but it's still present in the life force that we constantly receive as we arise in the consciousness of the Creator. If we search for it, we will become aware of it, along with the inner sound of the plane of consciousness that we occupy in our presence.

The Magic of Living in High Vibrations

Because we live in a plasma field of swirling patterns of electromagnetic waves, our consciousness interprets a frequency band that we recognize as our world of experience. In its essence, it is all energy. There is nothing solid about it, except in our belief system. By recognizing our chosen energies, which possess polarity, frequency or wave-length and amplitude, we are able to interact with them in creative ways. Emotionally, we can feel them. Intuitionally, we can know them. Our consciousness presents them to us in forms and sense impressions. When we encounter one another, each of us has an energetic signature that others can feel, just as we feel their presence.

Energetically, we have potentially unlimited awareness, because we participate in consciousness that envelops and interpenetrates everything. Our consciousness modulates energy. With our free-will intentions, we can pay attention to anything we choose. We can do this with our eyes and ears and all of our senses, and we can do it in our imagination and emotions, apart from the experiential world. If we let our heart guide our emotions and our intentions, we become positive in polarity, and we align ourselves with the unconditional love flowing to us in our life force.

As we elevate our awareness by intentionally living in joy and compassion, our consciousness begins to transform our experiences. Without the life force of our alignment with negative energies in either resistance or support, the negative force disappears from our lives. We can give our attention to positive energies. We all know when we're doing this, and we can feel it.

Because we have limited our awareness, we have not known how powerful our attention is. How we direct our awareness determines the level of our vibrations, which radiate the quality of our energy into the quantum field for manifestation in our experience. Whether we live in imaginary or empirical experiences, they're all the same in creating a vibratory level in our consciousness. By intentionally directing our realizations and emotions in alignment with the energy of our heart, we enter the world of magic, and our experiences transform into joy and ecstasy. The world around us, which is actually within our consciousness, adjusts itself to provide the qualities that align with our own vibrations.

Our personal experiences depend upon our own perspective and state of being. There are infinite positive energetic patterns and experiences that we can be aware of, and we are aware of as much as our self-imposed limitations allow. As we resolve our limitations and become our clear presence of awareness, vibrating in resonance with our heart, everything comes into alignment with us in our experience.

Transforming from Fear and Suffering into Infinite Being

Once we resolve our limiting beliefs about ourselves, we can be as expansive as we desire. In our true nature we are eternal, infinite Being, living in unconditional love and the ecstasy of limitless creativity. Except for our own self-imposed limitations, nothing keeps us from realizing the truth of our infinite awareness. We have accepted shame and abuse practically since birth. We have identified with the limitations believed to be real in our culture, including sinfulness, lack of love, imperfection and mortality. These are all artificial constructs that we have imposed upon ourselves in a kind of hypnotic trance, in order to experience what it is like to feel separate from our infinite Source. This has enabled us to develop our spiritual focus in ways not otherwise possible, but once we have gone as far as we can in the realm of duality, it is time to remember the Source of our life, and to awaken to our true, infinite Self.

In our physical embodiment we have been gifted with the ability to modulate and play with energy in unique ways and to exercise our creative ability in the empirical world. While living in the forms of physicality, we can be creative without delving into negative energy, but we have been fascinated with dark forces and what is possible in this kind of density. Now we know what it is like, and we can choose to live in the spectrum of energetics that we really want to experience.

We are not created to be enslaved in darkness. Suffering is

purely a choice on our part. It may have resulted from accession to subterfuge and dissimulation, but here we are. Because we have accepted fear into our being, and the belief in our mortality, we have enslaved ourselves into submission to threats of all kinds. We are not required to experience any of this, and we can transform our experience with a strong intention at any moment.

Because we express ourselves in terms of electromagnetic wave patterns, what we experience is a result of our own polarity and vibratory frequency. We control our energetic expression with our imagination and emotions, which results in a continuously changing energetic signature, radiating our state of being into the enveloping quantum field of all potentialities, and resulting in our attracting experiences that resonate with our own energies. We have absolute control of our own levels of vibration in what we pay attention to and how we feel about it and about ourselves.

If we could just erase our attachment to every negative experience we've ever had and focus our attention only on the kind of experiences we want, we could transform our lives immediately, but we have a lot of mental and emotional inertia, which holds us in remembrance of past negative experiences. These must be resolved by awareness, acceptance and forgiveness of ourselves for getting stuck in fear. The more we can imagine our true nature and feel what it is like to be unconditionally loving and loved by all of life, the more we can resolve the limitations of our ego-consciousness and transform ourselves beyond the realm of dark forces, which depend upon our energetic alignment for their existence.

Our Personal Hero's Journey

Opening our awareness to the expansiveness of the radiance of the heart of our Being, allows us to participate knowingly in universal consciousness. This is unlimited awareness, which our ego-consciousness is incapable of. To access it, we must intentionally raise our positive vibrations as high as we can. In every moment, this is a personal choice. If we can open our awareness to the full spectrum of our heart's energies, which we can sense intuitively, and we can align with them, we transform our lives.

As we vibrate higher in positive frequencies, we attract the same level of energies. We radiate the polarity and vibratory level of our thoughts and emotions into the quantum field, which responds by returning the same level of energetic patterns back to us as our personal experiences. If we can stay in the perspective of love and compassion, while also being emotionally and mentally clear, we experience magic and fulfillment. Our inner experiences can transcend time and space with an expanded awareness of reality.

Our limited view of ourselves disappears, because it becomes unbelievable. We can be present in our awareness throughout the cosmos and can create anything anywhere by knowing how to work with energy and being able to completely control our thoughts and emotions. We can live far beyond and include our body and ego-consciousness. When we are without limitations, our ego-consciousness dissolves. We are our present awareness, unfettered and free in every way.

By developing deep awareness of intuitive guidance, we can always know how to deal with any energies that we encounter. To be able to stay in positivity, we can maintain a perspective of gratitude, love and compassion. If we deviate from this, we discover our personal limitations that we can recognize, resolve and release, giving ourselves greater freedom. As we become more intuitive and align ourselves with its vibrations, we elevate our lives. We can learn to be completely Self-sufficient in every way, experiencing all of our heart's desires. By the high level of our vibrations, all of our actions and interactions can have the energy of life-enhancement for all conscious beings.

In our essence, we are pure Self-Conscious awareness of our

presence of being, unlimited in every way and having infinite creative ability. To become aware of our true Self is the goal of the inward path. Each of us can accomplish this with strong intention. It is our own hero's journey to the higher dimension of being.

Realizing a More Fulfilling Life

If we want to have an easy life, with an opportunity to pursue our passions with all the power we desire, we can choose to align ourselves with the unconditional love of our true heart. In this quality of energy we contribute to the enhancement of all conscious beings. Quantum physics has shown that there is a universal consciousness, which we all participate in. Even though our human experience comes from a different perspective, within the limits of our beliefs about reality, we can begin to realize that we all arise and live within the same consciousness.

Whatever perspective we desire to have is ours to choose. None is better than another, they just result in different experiences, which resonate with our perspective. Most perspectives have limitations, because they have a defined band of polarity and frequency that fills their state of being. Energies beyond this range are not perceived, but are still present. Only by being mentally and emotionally clear can we have a clear, expansive perspective, while we follow the guidance of our inner knowing, which is always positive and feels correct.

As our energetic environment becomes more positive and emotionally warmer, we can feel drawn into personal expansion of our awareness, resulting in greater wisdom and compassion. Our intuitive abilities are increasing as well. If we pay attention to these things, we can enhance the natural process of our spiritual evolution by intentionally aligning ourselves with the most positive and life-enhancing visions and feelings that we can create in any moment, sometimes by intentionally seeing the light in another person or situation and reinterpreting our encounters.

We love to have fun with one another, and we enjoy heartfelt relationships. Having these experiences begins within ourselves. It is how we feel about ourselves. In every respect, this is a choice on our part. If we have deep-seated emotional scars and knots, we can become aware of them by intending to, and recognizing their energy when they arise. We can realize that we have attachments to these experiences, but they are just experiences that contribute to our wisdom and compassion from having a victim consciousness and experiencing fear. We can resolve our attachments and choose a positive focus of attention, while being aware of negative energies that arise in our awareness.

By focusing on our inner state of being, we can realize our conscious life stream of positive energies, which we can align ourselves with. Staying with these energies ultimately leads to personal transcendence beyond the limitations of ego-consciousness, and to realization of our eternal presence of awareness.

Journey Toward Mastery of Self

As we grow in our understanding of life, we realize that we have lived within unnecessary limits, unless we have intentionally wanted consciousness limitation for specific kinds of experiences, such as living in the duality of the interplay of positive and negative energies. Apart from intentional, self-imposed limitation of what levels of energy we allow within our awareness, we are by nature infinite Beings, wielding powerful creative abilities. We only need to realize who we are as limitless awareness with free choice of focusing our attention on anything. We can imagine living in a realm that we totally love. When we realize this state of being, we align ourselves with its energetic patterns, and it opens our awareness to the reality of greater experiences.

With our vibratory alignment in resonance with an identifiable spectrum of energies, we create experiences for ourselves that are at the same vibratory level. Our experiences are a result of their attraction to our state of being. For this reason, we may seek clarity, so that we can more easily realize the reality of our creations. While we are attached to beliefs in our limitations, we cannot have clarity, and we tend to sabotage our creations unintentionally through doubt.

We have not believed that we are infinitely powerful creators. This is beyond ego-consciousness. If we do not release doubt in our ability, how could we realize our creations? Our power of realization depends upon our inner knowing of greater consciousness. True knowing comes through our intuition, which requires our unbiased, clear attention, allowing our awareness to be guided by the energetic patterns of the heart of our Being. This is life-enhancing energy in every way. It is our participation in universal consciousness expanding in unconditional love and joy.

By aligning ourselves in the best ways that we can with the most elevating thoughts and feelings that we can imagine, we can train ourselves to be open to the reality of living in a more positive and wonderful dimension of limitless potential for our creative intentions. As we work through our limitations, our awareness expands, until finally we realize that we are living in an ideal environment that is our present reality. The transformation of our reality happens when we can realize it, at which time any remaining sense of limitation becomes unbelievable.

Realizing Our Expansive Self

With the rising energetics of the Spirit of the Earth, we are being engulfed by positive vibrations in rising frequencies. It is becoming easier to be aware of the energy of the heart of our Being. As we become aware of it, we can learn to resonate in alignment with its expansiveness. Following it can take us beyond our limiting beliefs about ourselves. Our intuition leads us toward universal consciousness, where our awareness expands as far as we can allow toward infinite realization. Ego-consciousness can become our faithful servant, whom we can love. While we can be unattached to its limitations, it enables us to live in the empirical world as the personal expression that we create.

As we learn to use the power of our realization to create our reality, we keep bumping up against our limiting beliefs, which instill doubt about what we know. When we intend to be aware of the quality of energetics in every encounter, they can no longer control us. We can become Self-aware with expanded realization, having access to the consciousness that is everyone and everything. Once we realize that we are part of the one universal consciousness, the limitations that have bound our awareness begin to dissolve.

We can continue as our incarnated ego-consciousness, and we can regenerate ourselves on all levels. While we live in harmony with nature, we can regenerate our environment. This requires being only positive in loving confidence about what we are doing and creating. It all happens on a psychic level, as we connect with the quantum field, which manifests our experience of the energetic patterns we have created with our envisioning and feeling.

By asking for inspiration and paying attention to our intuition, we are guided to greater and greater realizations as we are able to accept their reality. This is a time of transcending our self-imposed limitations and realizing our true identity as aspects of our Creator, endowed with limitless abilities. As our awareness expands, our confidence grows, and we gain greater control of our mental and emotional abilities. Our increasingly expansive realization can enable us to be clear, present awareness with great creative power.

Using the Power of Our Realization

When we no longer engage with negativity, we stop giving it our life force through our realization of its reality. This frees us from its grip on us, while we feel enhanced by the surge in our life force, strengthening our personal radiance and brightening our essence. We can pay attention to all things positive, beginning with gratitude for our eternal presence of awareness. On this level beyond time and space, we live in ecstasy and complete fulfillment. We have been in this state of Being, and we can remember it, when we penetrate our consciousness.

At the same time, we can be aware of everything that is occurring around us in the time/space continuum. We can choose to participate in a positive way, with compassion and love, when we are personally faced with residue from our past dualistic state of consciousness. By maintaining a focus on the energy of our heart and realizing its reality, we radiate its loving, life-enhancing energy into the quantum field, drawing the energies of peace, unity and abundance into our environment.

By failing to engage actively with negativity, we allow it to dissolve from our experience into another dimension. From the ego-consciousness perspective, this is impossible. The ego does not recognize that we create our reality by our own realization. We are constantly creating our experiences by our realization of their reality. Realization in our awareness fills our thoughts and emotions, radiating these energetic patterns throughout universal consciousness and drawing our empirical experiences into alignment with our state of being, our level of vibration.

We can learn how to utilize our power of realization in transforming ourselves and all of humanity. In this process, we can use our imagination and our ability to call up emotions. Imagining ourselves as Beings of radiant light and feeling the presence of others as beings of radiant light, we can develop the ability to realize the reality of this scenario. We have experienced it prior to our incarnation, and we can remember it, when we intention-

Chapter 4. The Importance of Our Intentions

ally penetrate our consciousness. We can expand our awareness through directing and aligning our creative power of realization with our attention to the energetic level of our heart.

We live in the turning of the ages, and the energetic environment on our planet is shifting into a higher dimension. Humanity must make this shift into a realm beyond duality. The negative is dissolving as we replace it with our realization of a life-enhancing world. This is the shift we are making, and it is invigorating the Earth and regenerating everything.

Living in the Joy of Transformation

Our ego consciousness lives in the vibrations of duality and is constantly under stress physically, mentally and emotionally, and we are barely aware of our spiritual nature, if at all in our search for fulfillment. It is only when we move into a different expression of consciousness that we can feel fulfilled in every way. It is the expression of our multi-dimensional consciousness flowing through the heart of our Being and symbolized by our physical heart.

Once we get a hint that there is more to life than we have been experiencing, we can begin to open our awareness to greater consciousness. We can learn to align with the level of vibration that flows through our heart. This alignment can lead to a life of adventure in growing awareness and great expansion. It is living in the moment in resonance with our intuitive knowing. At the beginning of this quest, we do not know the larger picture of what is happening in our world, but as our sensitivities become more acute to our inner guidance, we can see through the drama and roll-playing that we all engage in.

We can become aware of our ability to create a miraculous life for ourselves and everyone who is open to our level of vibration. Our presence becomes life-enhancing in every way, filled with joy, serenity and laughter. We can be aware of everything happening around us and in the larger expressions of humanity and beyond. As we learn to transcend our limitations, we gain the ability to understand our connection with all conscious beings. Without being intrusive, we can be supportive of life-enhancement for all.

It is our realization of the reality of divine guidance in every moment that transforms our lives. We always have the choice to be aware of it, and we are free to create whatever experiences we desire, when we are in alignment with the heart of our Being. This is where we can discover the depth and extent of our presence of awareness. Since the consciousness that all conscious beings share has no bounds, our realization does not have to be limited to the realm of duality. By being intentionally positive in every moment, we can open ourselves to a realm beyond polarity, where the vibrations are always high in alignment with higher consciousness.

This level of awareness can enable us to live in the same world with duality and be unaffected by the negative. Being embodied with ego consciousness has inherent limitations that we can transcend by paying attention to the vibrations of our intuition. When we can remember our infinite Being, we are no longer limited within the dualistic empirical world. We become a transformative radiant Being with unlimited creative powers.

Guidance for Awakening to Greater Realization

The process of enlightenment is essentially a process of remembering. In order to penetrate the mind-wipe that we subjected ourselves to upon incarnating, we need to realize the truth about our self-imposed limiting beliefs about ourselves and the nature of ego-consciousness. The primary concern can be recognized in being able to feel the difference, even very slightly, between negativity and positivity. They are separated by a point out of time, with fear and doubt on one side and on the other are love

Chapter 4. The Importance of Our Intentions

and joy. In our state of duality, we participate in both, but not to the extreme of positivity. To reach that vibratory level, we can intentionally transcend our limitations and just be our presence of awareness. Intentionally without concerns, expectations or attachment to limiting anchors in the dualistic vibratory spectrum of humanity, we can achieve mental and emotional clarity.

In a state of clarity, we can understand the nature of the human experience, its constituent vibrations, and our process of awakening. We can understand and feel the quality of the range of experiences that are possible in this dimension, and we can know clarity without limits. When we are able to intentionally align with clarity of Being, we can be intimately aware of our intuition, which operates on the vibratory level of unconditional love and compassion. It is the light on our path to remembering our true essence, and it is drawing us into greater expansion into infinite awareness.

Nothing in the universe can resist our power of choice in our creative expressions. How we are in our mental and emotional state of being determines the quality of energy that we are radiating into the quantum field. Usually humans continue to create the state of being that they have accustomed themselves to. To escape this kind of energetic filtration, we can use our imagination. Our vibratory radiation is the same for our imaginary scenarios as our empirical experiences. We can learn to use our ability to create visionary experiences that we can feel ourselves living in. This is the quality of energy that we send out for the creation of the quality of our experiences.

Once we understand how we constantly create the quality of our experiences, and we accept personal responsibility for all of our creations, we can learn to be selective with our attention and emotional alignment. If we can allow ourselves to experience our eternal presence of awareness, we can open ourselves to the truth of our Being, our Self-Realization. This requires an awareness of that point out of time between polarities, the point of clarity, of being the presence of awareness of our infinite and eternal essence of Being with the free choice to create whatever we choose to recognize and realize as real.

Realizing Greater Consciousness

We can use all of our abilities to practice being our True Selves. Our power of discernment is critical in sharpening our awareness of the subtle energies separating negativity from positivity. Our doubt and fear have disempowered us from actualizing our creative manifesting ability. When we can be aware of the slightest tinge of doubt about our Infinite Being and creative power, we are very close to complete alignment with our true Essence.

When we can intentionally align with our inner vision and intuitive guidance in every moment, we can expand our awareness into universal consciousness. As more us achieve this state of Being, it becomes easier for everyone to align with the higher spectrum of vibrations in the realm of gratitude, love, compassion and joy. By imagining ourselves to be unfettered from limiting beliefs about ourselves, we can expand our awareness. At first this may be just a daydream, but the more we open ourselves to our inner knowing, the clearer it becomes, until we find that it is always present, prompting us to pay attention to the highest vibrations in the moment.

Being aware of how we feel in every moment enables us to be completely Self-guided in every situation. Knowing the vibratory quality of our intuitive guidance gives us understanding of everything that is happening from a perspective beyond ego-consciousness. In this state of clarity, we can modulate the energies in our presence to align with the vibrations in the heart of our Being. This is the process of transforming ourselves, humanity and all conscious beings into a state of enhanced living in greater vitality and joy.

When we are mentally and emotionally clear, we are powerful creators, projecting our full life force into the quantum field for manifestation in absolute confidence. With our intentional awareness of the light in every being, our natural radiance raises the vibrations of everything and everyone around us. Our life becomes miraculous and free. By choosing to live in unconditional love and life-enhancing energies, we transform our life experience into a high-energetic dimension beyond the reach of negativity. We can live in the spectrum of gratitude, compassion, love and joy, while also interacting with others who are living in the dimension of duality. In these relationships, we can feel within ourselves how to handle every interaction in confidence and in elevating ways.

Creating Miraculous Experiences

When we are able to control our mental processes and direct our emotions at will, we become powerful creators, able to project our full life force into our personal energetic expression. Without confining beliefs or personal needs and addictions, we can just be our present awareness. From this mental and emotional space of zero-point awareness, we can receive reliable guidance from our inner knowing, which is our awareness within universal consciousness.

As we become balanced and fulfilled through following our inner guidance, we can maintain a constant attitude of enhancement of everything and everyone in our awareness. We can learn to be aware of the presence of Creator consciousness in every entity. When we have an agenda that is open to whatever arises in our inner knowing, this quality of awareness becomes real for us. We can intentionally create visions and feelings that are brighter and more loving than we would otherwise perceive without knowing the inner essence of consciousness. By imagining with gratitude everything we are aware of to be beautiful and supportive, we open our awareness to those positive expressions and project our creative energy into their manifestation.

By intentionally expressing the energy that we feel within the heart of our Being, we are able to channel a great abundance of conscious life force through our powerful radiance, which may be unseen by most people, but is nevertheless transforming energies all around us. We become the expressions of unconditional love and enhancement of all life. In every moment, we can be in a positive mind-set and emotional state, with zero personal drama. When we can hold and wield our conscious life force in ways that align with our heart, we become fulfilled in every way, because the natural state of all life is an expression of Creator consciousness.

By resolving the negativity that we have expressed and lived with and have believed to be outside of our control, we can transform everything. The world that we experience, is the world that we have consciously aligned with. Its expressions change in synchronicity with our own vibrations. As we are able to maintain a positive state of being in gratitude, joy and compassion, we attract these qualities of energetics into our experiences and all around us. We become energy transformers, raising the frequency and amplitude of positive energy and empowering the manifestation of freedom and abundance everywhere in our awareness.

Accessing Guidance beyond the Mind

When we are aligned with the energy of our heart, we live in the moment. We are in the right place at the right time in every moment, always knowing in depth what is happening and where the energy that flows through us is going. We can live in the light of creative love and gratitude in every encounter, enhancing and enriching the energies around us, experiencing joy. Our past and our future are only potential experiences that depend upon our state of being in the present moment, and in moment-to-moment.

Chapter 4. The Importance of Our Intentions

In alignment with our heart-energy, time is irrelevant, because we always know where to be and what to think, feel, do and say in each moment. The moments can be a constant flow of gratitude, compassion, joy and love, and they can be filled with the flow of visions and everything that our heart desires, and everything that we enjoy doing and experiencing in ways that are life-enhancing for all.

We can learn to be very sensitive to negative feelings in us, and we can identify them as such and transform them. Impatience, along with all other negative vibrations, belongs to ego-consciousness. When we live in the moment, patience is not needed, because we know that we're always where we want to be. Until we know this, we can practice believing it. When we open ourselves to our intuitive knowing, and feel ourselves experiencing its energetic quality, it becomes our experience.

Eventually we can learn to be able to realize the reality of everything we can imagine and feel. This is what we are created to do. We are the creators of experiences to enrich universal consciousness, and we've been given the gift of deciding what kind of energies we want for ourselves. Because of our powerful training and programing in society, we have become deeply attached to our limiting beliefs about ourselves, not even realizing that they filter out of our awareness everything that vibrates beyond ego-consciousness. The ego has no higher guidance. It is oblivious to our inner knowing, because this is different from our rational mind. Inner knowing is beyond thought. It is the self-knowing truth arising in us through our conscious life force. It is infinite knowing within our awareness wherever we place our attention. This is beyond words to describe.

Living in the present moment is the only time. Our only awareness is here, now. By transcending our limitations and belief in time, we can intentionally open ourselves to our infinite presence of awareness in alignment with our inner knowing. Once we become aware of our real creative ability, our thoughts become things to play and have fun with, along with our emotions. No threats can impinge upon us, because we realize their nothingness without our attention and alignment. This is beyond belief for ego-consciousness, but we can find it to be true in our experience. It requires absolute inner alignment for us.

5.

Understanding Our Expansive Self

Aligning with Our Expanded Self

On the path to conscious expansion, we have many higher-dimensional beings in our presence, offering their light and guidance for us, as soon as we desire it and open ourselves to receiving it through our intuition. We can sense their positive, high-vibratory energy and align ourselves with it in gratitude for its uplifting feeling. If we can completely open ourselves in confident love and joy, we can receive so much life-enhancing stimulation to stay in their level of polarity and vibrations. It is a level of wonder and magic for our limited awareness.

Both subconsciously and consciously we have awareness of our intuition. It can carry us beyond the limitations of our ego-consciousness. This may take some practice to accomplish. The requirement is being open and receptive to intuitive knowing, and wanting to align with it. It's helpful to be mentally and emotionally neutral and just present in awareness. It's possible to be in this state of consciousness as we go about our lives. Our openness and intention draw us into the attracting polarity of higher-dimensional energy, and our awareness can expand greatly.

As we become more sensitive to our intuition, we gain confidence that it is divine guidance in every moment. It is in alignment with the consciousness of the Creator, and is life-enhancing in every way. As we gain familiarity with our higher-dimensional coaches, we can recognize what they feel like in the energy of their presence, just as we can feel our intuition. Because it is filled with vitality-enhancing energies, our higher guidance always feels good. Our natural energies are all positive and loving. In our true Self, we cannot know fear, because we are eternal, Self-Aware consciousness personified, and infinitely powerful creators.

In our true, expanded Self, we can create manifestations of ourselves in any dimension. Here we are expressing ourselves as human persons with limited conscious awareness in a realm of duality, which we have kept ourselves in, by creating limiting beliefs about ourselves. We are experience creators for universal consciousness. We are accomplishing the nearly-impossible by living in a deeply negative realm of duality and transforming it. We're doing it by resolving and transcending the limiting beliefs about ourselves, as we align with our intuitive knowing.

Whenever we want, we can return to our unlimited consciousness by making it our intention and following our intuitive guidance. Through our openness and receptivity and the desire of our heart, we can come into awareness of our eternal Being. We can be led to everything we need to come into alignment with our infinite Self-Realized Present Awareness.

Exploring Our Identity

We are our true Being with the human additions of personal limitations. Without the limitations, we are our eternal present awareness without limits, our expanded conscious awareness beyond the empirical realm of duality. Without personal, limiting beliefs, the ego disappears. We have created our limited human persons, but our true awareness and creative ability are far greater. We are designed to create all of our experiences through our modulation of energetics with the vibrations of our thoughts and feelings. Every personal belief has a vibration that binds us to it by its polarity and vibratory pattern. Whenever we want to expand our awareness into living in a kind and beautiful realm, our beliefs may allow only a split-second flash of its vibratory level, before it's dialed back for our conscious awareness.

The ego-consciousness cannot imagine the wonders of a higher dimension in awareness beyond time and space, but we can know its energy intuitively, if we are open and receptive to it. It is who we truly are as the creators of our human persons seeking certain experiences. We control the quality of energetics in our awareness by our attention, polarity alignment and resonance. How we feel and what we think about are our experience creators.

By intentionally being thankful and kind-hearted to all beings, we can align with the process of expanding our awareness. We can intentionally be completely positive in every moment. This is a multi-dimensional move on our part and involves our subconscious innate being. We must come into alignment with our subconscious self by being compassionate and understanding with ourselves, and by being open and receptive to realizing unconditional love.

Once we leave the realm of doubt and fear, everything becomes possible, because we do not uncreate our creations with our doubt. We can leave the realm of duality by aligning with the energetic level of being positive, compassionate and loving as much as possible and ultimately in every moment. Circumstances arrange themselves to resonate with our level of vibration.

Because we are enlivened eternally by our consciously creative life force, we are learning to use our abilities wisely. Everything we feel, think about and do can be life-enhancing all around and within us. As we train ourselves to live in positive vibrations, we find that we can be trustworthy to ourselves, accepting every situation with gratitude and openness, expecting in complete confidence an experience of joy, love, abundance and freedom.

Fractals of the Infinite One

Let's examine what it means to be a fractal. The earliest reference was in Indra's Net in the Athara Veda about 3,000 years ago. This was an infinitely large and complex design embodying the interrelated connections of all beings in repeating relations of mathematical precision, keeping everything in an energetic balance. Each node of the net contains a jewel that reflects all of the others. Indra's Net consists of fractals, each of which consists of fractals, which consist of fractals, all comprising an entire cosmos, with each jewel being a clear transmitter and reflector of its unique energy in resonance with all the others.

We are not just humans. We are the jewels in Indra's Net. We receive our connection with the infinite through the conscious life force that constantly enlivens us through the connection of our heart. We then modulate that energy with our thoughts and feelings, forming our unique energetic signature, which we radiate into the quantum field. In Indra's Net, the quantum field is symbolized by the spaces between the strands of the net.

MC Escher's graphic designs are a great example of how fractals appear and interrelate mathematically. This is how energetic patterns change and evolve with our awareness. Beyond the visual, fractals embody an essence that is shared by all and is part of the consciousness creating everything. By nature we are radiant, self-luminous jewels of many facets. Any doubt of our identity as divine Beings can be replaced by our greatest gratitude and joy in order to realize our connection to the Creator of all.

The Creator creates fractals of Self. That's who we are in our creative essence. We have chosen to be beggars, but we are actually masters who may not yet realize it. We have the ability to choose whom we believe we are at every moment. We just have to convince ourselves to drop our limitations. Our free will goes much deeper than is generally imagined, because it has been enslaved within our ego consciousness by our attachments, doubts and fears.

We can free our free will to be true to our fractalized creative nature by changing our polarity to positive and staying so. The qualities of our true Self are the same as the qualities of the Creator. This level of conscious vibrations is inherent in nature and the Earth herself. By aligning ourselves with the consciousness of our planet, we can open ourselves to our heart energy and our intuitive knowing, while clearing our vibrance as the jewels in Indra's Net.

Living a Positive, High-Vibratory Life

In our true Being, we are the divine Creator, fractals of the One. Although we have been largely unaware of our greater Self, we can open our awareness to our true Being. We can seek to know our unlimited Self-Awareness. Our inward journey is fraught with obstacles that we have placed there and have accepted as real. They are our self-limiting beliefs, which we adopted deep within our consciousness in order to have a genuine human experience in a realm of duality.

If we desire to move ourselves into a realm of positivity in every aspect of life, we will come into alignment with our natural life-enhancing energies. Our ego-conscious self cannot believe in its actuality. The realm in our consciousness that is beyond polarity does exist. It needs our recognition and openness for higher and higher vibratory visions and feelings. We can achieve a heightened state of Being by imagining and opening to Being in present awareness. Being void of attachments, random thoughts and feelings. Just Being present and open in eternal, unlimited awareness, guided by our intuitive knowing of everything we want to know in every moment. If this is our goal, then we must practice in order to get there.

To resolve our obstacles to expansion of awareness, we can recognize what they are and decide if we want them. This requires introspection and desire to know the unbiased truth about ourselves. We can learn to work in alignment with our innate being, which is not required to be hidden from our awareness. We only have believed it to be and made the separation real for ourselves. We have been intuitively trained to make ourselves unintuitive by creating our ego-consciousness. Our first step in expansion can be to remove the separation from our innate self.

We can communicate with our subconscious innate self by becoming clear in our awareness. Our innate self controls all of the functions of our bodies and makes our thoughts and feelings possible. We provide the awareness beyond all of that, the polarity, the choice of destiny. Our bodies can be regenerated by positive, high-vibratory living. They can express our state of consciousness, once we have moved beyond karma in our intuitive alignment.

When we are open, and mentally and emotionally clear, we can be completely aware of our intuitive guidance. All obstacles that may limit our awareness will have been resolved, and we can live in confidence, gratitude and joy.

Increasing Our Self-Realization

Because everything we experience is within our own consciousness, our experiences are a reflection of our state of being, the vibrations of our predominant thoughts and feelings. We are actors creating our scenarios by the kinds of energetic patterns and characters we attract by our polarity and vibratory level. We establish a caste with whom to play our roles for ourselves in conjunction with resonantly-vibrating others. Until we decide to explore beyond our limitations, we relive the vibratory levels of our historical experiences within the belief structure of humanity.

Our ego consciousness cannot go beyond our limiting beliefs without an unlimiting experience. It lives within the boundaries of our limitations and claims authority in this spectrum of energy. Our ego keeps us occupied and distracted by fear and stress in order to have our life force enabling its existence. Once we pierce the veil of our self-limitations, the ego dissolves, and we can realize our true personhood. We can become aware that we are One that can express our Self as our unique individuality in any dimension and as any kind of entity.

Currently we are humans on the Earth, but in our consciousness we can become aware of our greater Being. To open ourselves for this awareness, we can align with the positivity of high-vibratory emotional and visionary energies. In our true Being, we inhabit a state of Being that is entirely joyful and ecstatic in a realm of beauty and life-enhancing energies. This is what we can align with as much as possible in our imagination and feelings. The more we practice, the more vivid and convincing our experiences become, until they become real for us.

Identifying with our greater Self enables us to be loving and compassionate in our human interactions. Once we transcend ego-consciousness, we can take control of our human presence with the guidance of intuitive knowing. This transforms our lives into experiences that enhance all of life. We have the ability to create everything we need and desire with a pure heart. If we begin to live without limits, we find that we no longer have negative experiences, and we can know everything we want to know as our awareness expands into universal consciousness.

The Essence of Our Creative Nature

What we experience is what we have energetically aligned ourselves with. Every possible scenario exists in the quantum field as a collection of energies with specific polarity and frequency patterns. Although we believe ourselves to be solid physical beings living in a solid empirical world, our physicality is actually our conscious perception and interpretation of spinning energetic vortices of identifiable energetic patterns that become material when we recognize them. We recognize them when we align with their vibrations in our innate being. In our conscious awareness this means that we feel their vibrations and imagine their appearance.

We have the choice of aligning with any energetics in any dimension that we feel attracted to. By incarnating as humans on this planet, we have agreed to compartmentalize our conscious awareness within the empirical spectrum of energetics in a realm of duality. In order to do this, we have to be unaware of our greater Self, but we do not lose our connection with our true Being. This connection remains as our intuition in the energetic expression of the heart of our Being.

In our true nature we are multidimensional and are unlimited in every way. We are naturally creative in all of our thoughts and emotions, and we can learn to control and direct these aspects of ourselves for intentionally creative living. In order to create the desires of our heart, we must transcend our limiting beliefs about ourselves and become completely confident in our emotional and mental alignment with the energetic expression that we desire to experience.

Since everything we desire already exists in universal consciousness, we attract it into our experience by feeling its presence in our lives and imagining our enjoyment of it, until we recognize its reality in our personal experience. This requires no effort. We do not need to command anything. It requires only our gratitude and confidence in its reality in our experience beyond duality. If we can stay completely positive in confidence of our creative ability, we can create anything we desire through the energy of our heart.

In Search of Unconditional Love

In the deepest possible feeling. the experience of unconditional love comes to us in our inner knowing in every moment. To experience it in its true essence, we must know our Selves in our eternal, present awareness. Unconditional love never changes. It is our direct connection with the One divine consciousness.

In our ego-consciousness, we cannot know this. Our ego is designed to live in limitation in all areas of life, and we are attached to it and its limitations. It is unaware of higher guidance and expanded consciousness, but this is a voluntary attachment. In order to know our true potential in who we really are, we must transcend our ego-consciousness through opening ourselves to our intuitive knowing. The ego cannot interfere with this, only with how we receive it and recognize it.

Our intuition is connected with our inner senses, our innate being, and our conscience, and it is much more. It is our ever-present sense of knowing in the background of our awareness. By searching for this inner knowing, we can develop sensitivity to it by being open and objectively aware. We are prompted in every instance, just as we need to know, feel or do something. We may have other messages and vibrations coming into our awareness, but our intuitive promptings always come with the vibrations that feel life-enhancing in every way.

Nothing needs to be changed in our lives for us to experience the wonders of great love, joy and abundance, except our attitude and perspective. Instead of being judgmental and outraged about people and situations that we don't like, we can decide to be accepting, forgiving and kind, while imagining the inner conscious light in everyone and everything, and then interfacing with it. When we are truly aligned with the enhancement of all life, it will appear. This is the energy of unconditional love, and it is inherent in our essential Being. It is always present and never changes.

Unconditional love is the essence of universal consciousness. It is the nature of the Creator of all and arises in us through the conscious life force that we receive in every moment. It is the nature of our participation in the consciousness that envelops everything that exists in all dimensions. It naturally brings fulfillment to us of our every need and wish and much more.

To enjoy our awareness of it, we must resolve our attachments to the beliefs in our limitation. We needed our limitations for a genuine human experience, but our limiting beliefs are the only restrictions to realizing our true Being in greatest love.

Creating Our Reality

On the inward journey to enlightenment, the most important process to learn is how to raise our vibratory presence and become comfortable and consistent in moving into higher and higher positive vibrations. Since we have been created and have lived in higher dimensions, we intuitively know what they feel like, even though we have been unable to align with them in our compartmentalized human consciousness. We come from Being in unconditional love, light and truth in the fullness of Being, but in our human awareness, our ego-consciousness, we have been unable to know this. Yet, this is our origin and our destiny. We are here only for an excursion into the experience of duality, made possible by our acceptance of limited consciousness.

In our limited state of consciousness, we have been challenged to believe in the possibility that powerful, unlimited love in every aspect of life could be real. In order for this reality to come into our experience, we must transcend our limited perspective by sensitizing ourselves to our intuitive knowing through inner searching, when we are in a positive, high-vibratory environment in a state of gratitude. We can train our ego-consciousness to relax and just pay attention without desiring or thinking about anything. One way of experiencing unconditional love in our heart and emotions is to desire to align with it.

If we can train ourselves to be present in awareness and to feel the energetic presence of everyone and everything that comes into our presence, we can be aware of our intuitive guidance in every moment, and we can let ourselves feel what our heart wants to express for us. Aligning with this vibration and feeling carries us into a realm beyond polarity, where there is only unlimited love in everyone and everything. Here everyone is aware of the awareness of everyone that they focus on, and every focus and alignment is a sharing of life force and the energy of our heart.

Gradually we can move into universal consciousness, when we know deeply that our human self is a limited expression of our real Being, and that we are playing a kind of personal drama with predestined roles that we can change extemporaneously by intentionally directing our vibrations. In our essence, we are the ultimate creators of the vibratory level of our energy signature. This radiates our vibratory level into the quantum field, attracting resonating energy patterns and modulating them with our own vibratory patterns. This means that the vibratory level of our thoughts and feelings determines the quality of our experiences.

By being clear and without limiting attachments, we can

expand our awareness into universal consciousness and know everything we want to know in every circumstance. Our intuition is our connection with every conscious entity through the consciousness of the Being within Whom we live and have eternal awareness and our infinite creative ability. Our expressions come into our experience, when we realize them and feel the vibrations of their reality.

Working with the Creative Energies of Life

Although it appears and feels as if we are subject to actions, events and circumstances outside of our own being, we are absolutely in control of how these things affect us and even what kind of encounters we experience. We have learned to believe that we are our physical bodies, living in an empirical world that we can be aware of through our physical senses, and that we are subject to these external influences and forces. Anything beyond our empirical and mental perceptions, we do not know or feel.

As a result of the development of quantum physics just over a hundred years ago, we now know that there is much more involved in understanding the nature of our experiences. The quantum sciences have shown the validity of ancient spiritual teachings, that there is a universal consciousness that is the source and cause of everything that exists. There is only One Consciousness, and we all participate in it to the extent that our personal limiting beliefs about ourselves allow. Universal consciousness pervades everything and provides the essence for everything that exists. Everything is conscious, down to the infinitely minute subatomic constituents of our reality.

What does all of this mean for our personal experiences as we live our daily lives? Our conscious awareness consists of a fractalized essence of the Being that expresses Itself as us. Universal consciousness expresses itself in the form of electromagnetic waves that become material and experiential for us, when we recognize them. Without our recognition, everything remains imperceptible in the unified quantum field of all potentialities. When we can transcend our personal limitations, our awareness opens to a more expansive understanding of our reality.

As we become able to transcend our physical presence, we find that we are pure self-conscious beings of present awareness. We can avail ourselves of various techniques of self-transcendence beyond attachment to our bodies, and we can come into realization of our unexpressed, unlimited consciousness, including all of the emotions and visions of infinitely powerful creativity in the life-enhancement of all conscious beings.

We express ourselves as energetic patterns of electromagnetic waves with polarity, vibratory resonance and strength. What matters is the quality of energy that we choose to engage and align with in either support, resistance or transformation. Our inherent creative ability enables us to experience anything we choose, depending upon the polarity and vibratory level of our conscious perspective. Once we intend to align with the vibratory level of our heart, we can enter into alignment with the universal consciousness of the Creator of all, in unconditional love and expanding awareness. We can express our unlimited ability to create the fulfillment of every desire of our heart, and we can intentionally exempt ourselves from negative experiences just by our unwavering heart-felt energetic alignment.

Living in Creator Consciousness

Of the 26,000-year precessional cycles of the Earth, each cycle has a predominant characteristic level of energetics. For the cycle that we departed from on December 21, 2012, the polarity was largely negative, allowing for depression of human consciousness. It left a feeling of separation and diminishing life force. The cycle that we have now entered is largely positive, providing for more vitality and goodness. These energies are becoming obvious, bringing light upon the deepest, darkest energies of humanity for their resolution and release. Releasing the dark energies frees us from the bonds of enslavement that we have been under for millennia. This release happens as we realize that our enslavement has been possible only by our belief in its reality. This limiting belief is a choice for us to align ourselves with. We now also have the choice of changing our focus and aligning with freedom and unlimited awareness in unconditional love and gratitude.

If we choose to be unlimited, beyond time and space, we can open our awareness to our true, infinite Self, living in the presence of the Creator of all. In this state of Being, we align with the vibrations of infinitely creative energy and enhancement of all conscious beings. We all exist within the universal consciousness of the Creator, in which we can open ourselves to the awareness of any conscious being and be telepathically connected, once we become sensitive enough to our intuition. This is an intentional process of resolving and clearing our conscious and subconscious blocks to our connection with our higher guidance.

In the flow of increasingly positive, high-vibratory cosmic energy, we are being carried toward Self-Realization of our expanded Being, beyond our human experience. As fractals of the consciousness of One, each of us is an infinitely powerful creator in our true Being. In our human being, we have chosen to assume limiting beliefs in order to participate in this game of consciousness. From a more expanded perspective, we can understand its purpose, and we can change the quality of our experiences, which may also change the place and form of our lives.

In this life we are learning how to control the flow and direction of our life force. The flow of our energetic radiance results from our level of expanded consciousness, resulting in our radiance of joy and life-enhancing energies. We can learn to be present in our awareness as we go about our lives, and we can be aware that we are the presence of Creator Consciousness in every moment, expressing Itself in everyone and everything that is happening. When we realize that this is our reality, it immediately comes into our experience.

Aligning with Our Infinite Self

When we are attached to persons or moments of negative or positive experiences, in either case, we dwell on them with our attention, and we align with their vibratory levels. This keeps us from paying attention to greater ways of being. Until we free our attention, we're stuck in a limited experience within duality, with which we've become accustomed and comfortable. If we're stuck in a negative scenario, it may seem as if there's no escape. But there always is. It is our power of imagination and focus of attention.

Through deep understanding, we can create the resolution of our limitations. If we can recognize that we create and maintain them, we can accept our situation with compassion for our ego-consciousness, who has directed our energetic expressions without higher guidance. We can then change our focus and put our attention and energetic alignment on more elevating visions and feelings, and we can continue going higher for as long as we desire. In our natural, limitless state of being, our awareness is unlimited. This is where we're going in our Self-Realization as beings of light and eternal love with infinite creative ability.

We always have the choice of being at our chosen level of vibrations, regardless of what kind of energy we may be interacting with. Once we can release our attachment to limiting beliefs, we gain the freedom to expand our vibratory alignment into more positive and elevating vibrations in our imagination and feelings. By having no mental or emotional attachments, we open ourselves to realization of our expanding awareness.

In our unattached state, we're not creating the energetic patterns that we've already experienced. This enables us to choose a higher focus in each moment, because our energetic level creates the quality of our experiences. By being only positive at the level of compassion and gratitude, we can move closer into alignment with the energy that enlivens us.

In the state of just Being our present awareness, we have limitless potential in creating the qualities of our experiences through the use of our attention and alignment. Our current situation in the world of humanity is immaterial to our creative ability and realization. Once we decide to take conscious control of our creative abilities, we become the directors of our attention and of our thoughts and emotions, and we can also be the directors of our subconscious. In this position, we can be aligned with our intuitive guidance and can live in the awareness of miracles and love all around us.

Knowing the Expansiveness of Our True Being

If we decide to be as expansive as possible, we may be aware of all the places where we're ethericaly stuck in negative patterns of energy, like emotional knots, that we can resolve and release, once we recognize them. As we release them, we gain stature and confidence in greater awareness of our own presence as well as the qualities of the presence of others.

Our physical bodies represent to us all of our limitations. Our focus on them causes them to affect us personally and also to affect how we perceive and relate to others. Any defects that we observe in ourselves and others are part of our perspective on life, and they control where we are constricted because of some kind of fear, doubt or anger. However we acquired these habits of being, they exist to keep us locked into a compartment in our consciousness, keeping us from awareness of the truth of our Being.

We exchange energy with others through our eyes as well as by our etheric presence. We are constantly emitting photons and radiating the energetics of our state of being. As we expand into a higher vibratory mental and emotional presence, we become more radiant and brighter. If we can let go of the importance of our physical bodies and just focus on the energetics of the presence of everyone in our awareness, we can realize that we are expanding in our awareness of the energies of life.

What we have believed is real has locked us into the limited consciousness of the empirical world and a dualistic state of being. By choosing to be only positive, we can empower our subconsciousness to regenerate our bodies through our alignment with the energy of the intuitive radiance of our heart in its etheric presence. If we can focus on this level of vibration, we can become aware of our expanding presence and develop an elevated perspective that interacts with compassion, gratitude and joy.

While we focus on the energetic presence of everyone and everything around us, we can learn to treat them with compassion and love. If their being is receptive to these energies, we can imagine them as beautiful and vibrant. And so they will be, when we recognize them as such. For us there is no consequence, whether our visions and inspiration are received by anyone else, because we are creating the vibratory level of our own experiences with our etheric support of others. It is our own rising and expanding vibratory level that is important for us and for everyone else. We can fill the ethers with the vibrations of vitality and life-enhancement.

How Unlimited Can We Be?

We have a natural desire to be infinite in our awareness, because it is our true nature, beyond time and space, as well as in our human form. As humans we are focused on our empirical reality, but our awareness in not bound to the physical world. We can go deep in meditation and dreams, transcending our human hypnotic trance. We can choose to open our awareness to another dimension by using our imagination to identify patterns of energies beyond polarity, where all energy is positive and is emotionally identifiable as such.

As we invite our imagination to follow the energy of our heart into a higher realm, we can begin to recognize waves of love and compassion encompassing and interpenetrating our internal experiences. With our intentions, we can expand our awareness into wondrous scenarios, as great as we can believe are possible. When we recognize them, we can realize them, and they become real for us. As we learn to adapt to higher consciousness, we can feel and be more unlimited in our present awareness.

Our creative ability lies in our feeling, recognizing and aligning with the vibrations of any quality of energy. Our mental and emotional alignment with energetic patterns or scenarios makes them real in our experience. By living intentionally in gratitude, we can stay positive, while we create positive experiences for ourselves, regardless of what may be happening around us in the world. We can live in a sort of parallel dimension on a higher level, where dark energy is absent. It manifests for us according to our level of our confidence.

There is no power outside of our own consciousness that limits our creative ability, and we are the sole directors of the kind of experiences we draw out of the quantum field. If we pay attention to the fabricated world of humanity, we cannot find higher guidance for understanding life. The world of humanity is the realm of ego-consciousness in the experience of the duality of negative and positive energies. Transcendence of the ego through intuitive knowing is our connection to greater awareness.

We share our energetic presence with many beings, seen and unseen, and we are the sole directors of our experiences. Our quality of life is a result of how we think and feel about ourselves. Nothing outside of ourselves keeps us from being unlimited. We are invited to drop our attachments to our inferior beliefs about ourselves through compassionate acceptance, forgiveness, love

Chapter 5. Understanding Our Expansive Self

and transformation. As we elevate our perspective beyond our self-imposed limitations, we can be as expanded as we can realize.

Exploring the Essence of Who We Are

We have lived in our ego-consciousness for eons, as if this is our total reality. Although we have expressed ourselves as our individual human conscious self, this living being is only an extension of our true Self, our Great Spirit. Beyond our human self, we are Self-Realized infinite awareness. We are our limitless conscious presence, potentially aware of the awareness of all conscious beings everywhere. We have also lived without knowledge of our infinitely-powerful creative ability; however, there is nothing keeping us from knowing who we are and what we're capable of, except our self-imposed limiting beliefs about ourselves.

Our limiting beliefs are so deeply ingrained in our consciousness, that we have not even been aware that they may not be real. We have not imagined that we are immortal, because we do not have empirical proof of it. Of course, empirical proof is impossible, unless we die and return to our body, as some have done and reported their experiences of greatly-expanded awareness. Once we realize that we may be greater than we have believed, how do we resolve our limitations and transcend them?

The real question is how do we really know anything? We have beliefs about what is real, but what is their basis? Because of the way that our consciousness interprets the spectrum of energies that we recognize as empirical, we have believed that the empirical world is what is real. Now we know that it is all patterns of energy, and these are part of a much larger spectrum of energies that we keep ourselves from being aware of.

There is an aspect of our psyche that is unlimited, and that is available to our limited awareness. It is our intuition, and it is the source of everything that we truly know. Beyond what we have been taught, and what we have perceived, we have an inner knowing. It is different from what anyone else knows. It is our connection with our own infinite Being, and it knows everything about us.

In our true Self, we live in the consciousness of the Creator of all, and we share in the Being of the Creator. Once we transcend the limitations of the ego and its perspective of separation from the divine, we have unlimited creative power available to us. By our intentional sensitivity to our intuition, we can expand our Self-awareness. We can examine our limitations in the light of our intuitive knowing, and we can hold the intention to be so much more than we have been, by receiving intuitive awareness of the presence of unconditional love and joy in our lives.

The Depth of Our Being

By being in a state of deep gratitude, we can allow ourselves to be drawn into awareness of eternal Self-Realization. Once we have the awareness of our inner light and the awesomeness of our Being, we have transcended ego-consciousness and negativity. Our life transforms to experience the vibrations of love and fulfillment, regardless of what appears to be happening around us. Once we know our true Self, we can align our human self as an expression of our eternal Being. When we realize our infinitely powerful creative ability, the limitations of empirical life disappear.

Coming into Self Realization can happen by being intentionally completely positive, enhancing all of life in our thoughts and emotions. This is the expression of the energy of the heart of our Being, and it resonates in our intuition. By choosing to be in this energy spectrum, our experiences come into alignment with us. When we follow our inner guidance, we can achieve complete freedom in every way. We can resolve and release our limitations and all negative energy patterns, allowing us to transcend ego-consciousness.

In gratitude, we can imagine being our true Self, ever expanding our awareness in universal consciousness. We can keep clarifying our awareness, until we become transparent. In the consciousness of the Creator, we can realize the flow of divine energy surrounding us and allow it to draw us into alignment with positive, high vibrations.

As we move through our experiences, we can have the perspective of the Creator and send our thoughts and feelings to the healing and enhancement of all that we encounter. We can imagine positive transformation in negative situations and continue to feel gratitude in all experiences. By having only confidence in all our creative thoughts and feelings, the desires of our heart naturally manifest for us in ways that we do not need to imagine.

We can become the masters of our lives, able to create experiences that we love and cherish, because they are manifestations of our gratitude. We can be the transformers of humanity, as well as of ourselves. It happens naturally as a result of the polarity and level of vibration of our state of being, our mental and emotional energetic alignment. By our realization, we create the reality of our experiences.

Realizing Our Infinite Self

The Infinite One experiences the awareness of every conscious being everywhere, down to the smallest sub-atomic wave-particle being and out to the universes. Each of us is living in the life stream of our Creator and participating in universal consciousness. We have complete free will in choosing how we want to feel. As we confront the energetic patterns that are brought into our presence, our feelings are stimulated. Our emotions are both informative and creative, depending on how we use them.

Our reaction or acceptance of each encounter determines the quality of our experience. If we align with the energies, in order to resist or support them, we give them our life force through our recognition and engagement, making the experience real for us. If we accept the energies of our encounters, and we maintain our alignment with our infinite Self through our intuition, any negative energetic patterns will dissipate out of our presence. We no longer believe in their reality, because they have no basis in Creator consciousness. They are real only because they are human-created within duality.

As we complete our excursion into time-space and duality, we can learn to master the human experience with our creative intent. By aligning with the energies of life-enhancement for all, we can confidently create the fulfillment of the desires of our heart. We can thoroughly enjoy living on the Earth and sharing her energy. Being clear and having consciously and subconsciously no constraints upon our awareness, we are free to modulate all the energies in our presence in resonance with compassion, gratitude and joy.

The polarity and vibratory level of our presence is expressed by our perspective, and our alignment with our presence beyond time and space can give us a perspective beyond ego-consciousness. Once we are aligned with our intuitive knowing, we naturally transcend karma, because we no longer need its lessons.

If we intend to realize our unlimited Self, we can begin by imagining the most elevated version of ourselves, and continue to upgrade that as our realization becomes greater. Who we think we are depends upon our beliefs about ourselves. If we examine our personal beliefs very closely, we'll understand how we have limited ourselves, and we can decide how we want to be. Changing our beliefs changes our character and our energetic signature.

When we can realize our unlimited Self, we can free ourselves from beliefs in personal limitation, and our intuition becomes our constant guide to knowing and understanding every situation. As we intentionally align with the energies of life-enhancement, we align with the consciousness of the Infinite One. We are free to have fun and enjoy life, anticipating being loving and grateful in every moment.

Realizing the Transcendent Truth of Our Being

We have had the perspective that staying alive here is a constant struggle, interspersed with some good times. Do we have the ability to change ourselves enough to transform our experiences into the life we really want, beyond any negativity? Some people have special abilities that their consciousness can control, such as telepathy, empathy, clairsentience and many others. These give us glimpses beyond the empirical, and they are all part of the same consciousness that we participate in. If we haven't realized that we have them, we have blocked our awareness of them in order to have our own unique experience, because of something that we wanted to learn. Ultimately, spiritual learning leads us to the realization of who we are as fractals of universal consciousness.

Many spiritual masters have shown us the way to true knowing and Self-Realization. When we choose to pay attention to it, the way becomes clear for us, and it is unique to each of us. When we begin to recognize it, we can feel the warmth of attraction. We also have more advanced beings in our presence, but in higher dimensions, advising and helping us to achieve what we want. When we ask and are open to receiving their guidance, it comes immediately through our intuition. We also have our own intuitive knowing from our greater Self, which we can become aware of, when we open ourselves to it.

One aspect of our being that is the same for everyone is the realization of our intuitive knowing. By aligning with our intuition, we can resolve and transcend our limitations and come into realization of our infinite Self. Once we choose the path to greater awareness, we can begin to realize that this is our reality. Our intuition can guide us to the realization of unconditional love inherent in our Being and the life force that we share with all conscious beings, giving us the same essence of life.

As fractals of universal consciousness, we have an infinite ability to create, and we do it all the time, just by our vibratory level. Whatever mental and emotional quality we vibrate at is the quality of experience that we are creating. Universal consciousness envelops us, and we have access to as much of it as we allow. By living in positivity and high vibrations, we naturally attract energetic patterns that align with us. Our presence of awareness can be unlimited, as well as our creative ability. This happens when we can align with its vibrations. In expanding awareness, recognition and realization of our infinite Self can become our reality.

Awareness beyond Our Beliefs

As we can be aware of the awareness of others, including animals and plants, our Creator is also aware of our awareness. We are our Self-Aware presence, both embodied and energetic, beyond the physical. Our ego consciousness can be aware of a limited spectrum of energies, but does not realize its limitations. This limitation provides the reality of our empirical experiences. For us, it is so real that we have become fixated on it, unable to expand our awareness into our inner knowing.

Among quantum physicists, the Consciousness of the Creator is known as universal consciousness. It is the same consciousness, and it is the One that we participate in. We are created in universal consciousness and given infinite awareness. Although physicists are not aware of the array of qualities in universal consciousness, we can be aware of them through our intuition, just as we can be aware of the awareness of our pets. We can feel how they feel, and they can feel how we feel.

In universal consciousness there are no secrets. The quantum field holds everything always, and we choose how much of it we allow ourselves to be aware of. As we resolve our limitations, we resolve our ego-consciousness, and ultimately we can open our awareness into universal consciousness. We can keep expanding to infinity and also be self-aware of our human personhood. In our clear presence of awareness, we are multi-dimensional and can create our self-expression to be anyone we desire to be in any environment. We can transform our lives at any time.

It is our realization that creates our reality. When we realize and align with our intuitive guidance, we experience the results in our lives. Realization is beyond beliefs. It is based in inner knowing and can see through beliefs to recognize their basis in fear and limitation. Our intuition is our connection with universal consciousness, and it guides us toward Self-Realization. Once we realize our true essence, we can be fulfilled in every way, because we are infinitely powerful creators, fractals of the Creator of all.

We are embodied here now to transform ourselves into a higher dimension of energetics. This time in history is the turning of the ages. Love and beauty are replacing fear and ugliness. At the same time, the Earth is transforming into a more vibrant expression of higher vibrations. If we can release all attachment to limitations, and focus on our intuitive knowing in joy and compassion, we can align ourselves with the energetic level of our true Self. In our true Being, we can be masters of our lives, making real in our experiences the expressions of gratitude, love, joy, freedom and abundance. The consciousness of the Creator enhances all life. Through our intuition, we can align ourselves with this vibratory resonance.

Realizing, Embodying and Expressing the Truth of Our Being

In every now moment we are present as great Beings of Light, infinite in our awareness and creative ability, but until we can free our attachment from our limitations, our human aspect cannot realize this. All it requires is changing our perspective, releasing our limiting beliefs about ourselves, and aligning ourselves vibrationally with our true essence. We can take a few deep, rhythmic breaths and center ourselves in the energy of our heart. Here we can envision ourselves in the image of the Creator, as fractals of the divine One. We can feel inspired in the light, filled with vitality and unconditional love for all. We can envision every cell in our body as a perfect, radiant and regenerated expression of our greater Self.

By holding this vision whenever we can, we begin to realize its reality, and our life experiences begin to manifest this energy for us, regardless of what may otherwise be happening in our periphery. If we can hold the feeling of being filled with divine light, it transforms our human expression and experience. We become able to interact with the light in everyone we encounter, even those who have rejected their own, having stolen what they have from others through deception.

When we have thoroughly aligned our awareness with our inner knowing, we cannot be threatened, because we have released fear and belief in our mortality. We vibrate in a dimension beyond the reach of negatively-polarized beings, who have access to our consciousness only if we engage with their vibration.

Everything exists within our own consciousness, even if we don't realize it. If we limit our awareness, we pretend and believe that there is an entire world outside of us. The way that we can know what is true is to align with our intuitive knowing, because nothing beyond our own consciousness is provable. Everything is hearsay and appearance. What is real is universal consciousness and its expressions of infinite patterns of electromagnetic waves, some of which we perceive as empirical reality, because that's how our consciousness interprets them, according to our beliefs.

Once we begin to know ourselves as great Beings of Light and choose to vibrate in alignment with our inner knowing through the energy of our heart, we transform our lives into expressions of divine awareness and limitless, intentional creativity. This is the destiny that we are here to realize. We are eternal expressions of the divine One.

Opening Ourselves to Our Deepest Being

Our entrancement in our ego-conscious experience of life has been so complete, that we have been unaware of our greater Being. If we can extract ourselves from our electronics and all of the stimulation that takes our attention away from pure presence of awareness, we can be serene and at peace within ourselves. We can leave the realm of parasites and enter the realm of true love, where we are Self-Realized and completely fulfilled. This is possible for us, because of our creative consciousness.

By being present and clear, we can direct our attention to any level of energy that we feel attracted to and feel its essence. If we seek positive vibrations and begin to feel them in our awareness, we attract them into our experience. By intending to elevate our lives and those we encounter with joy and brightness, we receive inner guidance in how to do this. This intention takes us beyond ego-consciousness into a realm of heart-felt knowing.

With the increasingly higher vibrations of the Earth and our surrounding cosmos, we are being carried energetically into a higher way of being. An enveloping photon cloud is making everything brighter. Because photons are conscious beings, they seek every possible entry to places that have remained hidden from our awareness. Darkness is being illuminated and revealed everywhere, along with those who resonate with its negativity. As we change our focus and vibrate with the life-enhancing energy of our heart, the dark force is losing its essence. It cannot exist without our support.

Physicists know that higher-frequency energies are more powerful than lower-frequency energies. Because unconditional love is the highest frequency of life, it is the most powerful. The only reason that the dark force has seemed powerful is that we have allowed ourselves to be compromised with self-limiting beliefs about ourselves that have lowered our vibrations in alignment with the negative. It uses our life force to control us. Once we realize our situation, we can resolve our limiting beliefs and free ourselves to align with the desires of our heart and participate in a more powerful and enlightened way of being.

Because we are created in the power of unconditional love, we are great Beings of Light. In our essence we are present in our awareness beyond time and space and are infinitely powerful creators with the qualities of our state of being. With our energy signature we radiate our vibrations into the quantum field for manifestation into our experience. By intentionally vibrating in the dimension of creative love and joy, we release ourselves from negative experiences and open ourselves to fulfillment in every aspect of life.

Realizing Our Reality

Created in the consciousness of the Creator of all, we are fractals of universal consciousness, having all of the properties and capabilities of the One who originates us. We are our individual eternal present awareness, endowed with the capability of creating universes. We have always existed and will always exist in any form that we choose or in no form at all. The empirical world that we currently experience results from our interaction with the quantum field of all potentialities, which is an energetic expression of universal consciousness. In order to be aware of the empirical world, we have developed our ego-consciousness and imposed upon ourselves a set of limitations that function to make this world seem solid and physical by believing in its reality. We have become so entranced in this empirical world, that we have forgotten the capabilities of our greater, expanded Being.

Those of us who have experienced life beyond the body and the empirical world understand our eternal Self and expanded awareness, but this experience does not necessarily give us the perspective necessary to master the empirical world. We still have to realize our beliefs, how they function in our lives, and how they keep us enclosed within a compartment of our consciousness that we transcend when we are beyond the physical body.

In the process of incarnation, we designed a way to expand our awareness beyond our conscious compartmentalization. It begins with our intention to deepen our understanding of the truth of who we are. Intuitively, we know this, and we can intend to follow our inner guidance to self-revelation and remembrance of our true identity. We have been given much help by quantum physicists and masters of esoteric spirituality in understanding the workings of consciousness and our interaction with the quantum field and its energetics.

There is no such thing as solid physicality. At the subatomic level, what appears to be material is, upon technological analysis, swirling patterns of conscious entities that manifest as points of present consciousness unlimited by time and space. Time and space are constructs of our own consciousness that enable us to recognize the collection of subatomic beings as material. This world is constructed and deconstructed trillions of times each second, and we constantly interact with it in our mental and emotional state of being.

By our beliefs, imagination and feelings, mostly about ourselves, we create our own energy signature. In this way we interact with resonating energetics in the quantum field. Everything we could ever want is available to us, when we are in a constant state of gratitude for everything in our experience and when we are feeling fulfilled as the infinitely powerful creators that we are in our essential Being.

On the Way to Stardom

Everything that we could ever want is available to us now. All we have to do to experience it in our lives is to recognize it in our imagination and feelings and know that it is real for us. We create our experiences with the help of divine energetic plasma waves in the quantum field enveloping us. They are always responding to our visions and emotions. Through our intuition, they are non-intrusively guiding us to free ourselves from the compartment of consciousness that we have limited our awareness to. Once we know that we are constantly creating the qualities of our experiences just by the way we feel and think, we realize that we can live in deepest love and joy.

When we realize the power of our essential Being, with our unlimited creative abilities, we are completely free, no longer enslaved in duality. We can be aware of the nature of all the presuppositions that underlay our way of life, and we can question them. We may find that they are all fantasy. We have become so enmeshed in our self-imposed enslavement, that we have attracted negative enforcers in the form of our elite controllers, to whom we constantly give our life force, so that they can continue to steal from us on many levels.

The way out of a limited life is to transcend it in every way by adjusting our personal vibrations to be only positive in a high-frequency band. By having a perspective that is based in unconditional love, we modulate the energies that we encounter into resonance with us. We are no longer attracted to the clever, deceptive, life-diminishing energies of negativity. We can enhance the life of every conscious being we encounter just by our presence of awareness radiating love and compassion.

With the realization of our eternal presence of awareness, we can know that we can be absolutely trustworthy. This is an important achievement, because our ego-consciousness does not trust itself. It doubts all of our true abilities and is distracted by nearly everything. Not trusting ourselves results in the stifling of our intentional creativity, and so we have lived with our limited beliefs about ourselves.

There are things that we can do to bring ourselves into a state of serenity. We can listen to inspiring music, work with our breath, absorb the energies of natural environments and more. We can learn to sit quietly and imagine how we arise out of the consciousness of the Creator of all, along with everyone and everything that exists. We are all the same Being, sharing the same universal consciousness, all enveloped in unconditional love and joy. We can live in only positive energies in a dimension more divine. When we can absolutely trust ourselves, we become the masters of our lives, with infinite creative power, expressing the radiant energy of our heart.

Self-Transcendence of Ego-Consciousness

Today there is significant research being conducted on behalf of people who are psychically severely impaired. They're not criminals, but society cannot deal with them. The research has been very useful in helping the patients to remember their greater Self. The medicine being used is peyote, psilocybin mushrooms and ayahuasca. Their experiences with these medicines is carefully monitored and recorded. What happens? Their current understanding of themselves fades away as their awareness expands greatly and they have freedom to focus on whatever they are attracted to. For many, this is a novel experience, because they have felt under the control of other forces in their normal lives.

They find an inner voice or feeling that they recognize as truth. They feel really good, many for the first time.

They don't usually remember these experiences, but there is a subconscious effect that eventually dissolves the old programming of self-limitation. The point of all of this is to cure a social problem, but it has much wider implications. Medicines are just crutches for our greater consciousness to get our ego-consciousness to use in order to believe in a natural psychic process. We are able to have the same experiences by aligning with the same energies that the medicines emit. Under the guidance of nature, we can begin to vibrate in resonance with the energy of our heart and our intuitive knowing.

Through deep meditation, it is possible to do this. We can take this path by learning deep, rhythmic breathing exercises like Tantric breathing and some kinds of yoga breathing. We can also use other techniques of resolving mental and emotional interference in our clarity and serenity. Traditionally these techniques require much practice for many years in training the ego-consciousness to believe in a greater reality. Bio-feedback training can be just as intense. Medicines provide an accelerated path and may provide an eventual complete transformation, but if we can change our beliefs about ourselves intentionally, we can achieve a permanent transformation. Beginning with breathing and moving into serenity without inner interference in the focus of our attention, we can pay attention to the vibrations on the line between waking and sleeping. This is a completely relaxed state of pure presence of awareness.

In this state we are free from our normal human consciousness and our physical body sensations. We have transcended our limitations of ego-consciousness. We are in a sort of lucid dream of infinite Being. This is a completely different kind of realm and takes some getting acquainted with. Essentially we are exploring our own consciousness without limits. It is here that we can develop an unconditional relationship with our inner knowing.

Chapter 5. Understanding Our Expansive Self

Evaluating Our Destiny

One of the most significant proofs of quantum physics is the realization of universal consciousness. Everything that exists participates in universal consciousness and has no existence apart from it. Consciousness is the essence of everything and endows every living being with its vitality. There is only One unified consciousness, and all conscious beings participate in it. This realization eliminates the idea of a God who is separate from us, and it shows that our consciousness is the consciousness that creates everything. We may call it the consciousness of the Creator, because that's what it is. Although in our human consciousness we severely limit our awareness, we share in the consciousness of the Creator of all and have the potential of infinite awareness within universal consciousness.

At this most significant time in human history, we are receiving great assistance from our cosmic environment and from the Spirit of the Earth to transcend our limited awareness and awaken into universal consciousness. The resonant frequency of the Earth is rising and carrying every inhabitant with it. The energetic field enveloping us is becoming more positive and supportive of greater love and wisdom. Those who cannot adjust to these changing and elevating energies are becoming uncomfortable and will not be able to be here much longer. Those who are staunchly negative are becoming unstable and are going insane, as we can see with members of the ruling elite. Negativity can no longer be hidden and must face the light for transformation or demolecularization. This is happening all over the world.

Universal consciousness is only positive, but it allows for the operation of free will, and we have been allowed to create negativity and keep it active with our life force. Any kind of lack, threat or danger is a result of human creation. It cannot exist without our creative energy. Because each of us is the Creator, we have been creating our own negative experiences either intentionally or by acquiescence. Either way, our life force has been used through our attention and our conscious engagement, belief and alignment with negativity. We have the ability to change this by choosing to be only positive to align with the energy of our heart and to align with our intuitive knowing.

We are constantly created to participate in unconditional love and fulfillment in every way. We only need to realize this and choose to accept it as our reality. We have been misusing our inner powers to create what we don't want. It's all a matter of how we use our attention, what we choose to think about and how we choose to feel about ourselves as we live our lives. Within our own being, we have the abilities we need to transform our lives whenever we choose, and we are being prompted by the cosmos to do so now.

Making a Shift in Consciousness

By bringing ourselves to a state of serene equilibrium, we can make a shift in our consciousness to a higher dimension. By listening to our inner sound, we can calm our mind. It draws our attention inward, and we can begin to penetrate our awareness and realize that we are beyond our ego-consciousness. Our identity is so much greater, as are our creative abilities. This realization can open our awareness beyond our limiting beliefs and allow us to transcend them.

We can intend to enhance our lives by directing our attention to the positive spectrum of energies of gratitude and joy. These are the vibrations that can transform our lives. We can begin to be compassionate with everyone, including those we have regarded as evil. They also have a higher aspect of light within their consciousness, even though they have become parasitic. We need not engage with them on their level, as we withdraw our life force from them, and they will disappear from our experience, unless we invite them by aligning with their negativity, whether in resistance or support.

If we can intentionally direct our attention to positive vibrations in every moment, and interact with the light in everyone, we become positively radiant. We can accustom ourselves to being aware of knowing the energy of our heart. It is our higher guidance. In this way, we can be in the presence of our expanded Self, providing us with wisdom and inner knowing of all that we need and want.

This kind of shift in our consciousness transforms everything in our lives, because we enable ourselves to live in positive vibrations. This is the energy we radiate into the quantum field, attracting energetic patterns that resonate with us. Negativity becomes non-existent in our personal experience. It has no alignment with us and is not in our spectrum of vibrations.

In a positive state of being, we're able to have clarity of mind and emotion. With confidence, we can realize our creative intent in every moment. By maintaining confidence in our mental and emotional projections, nothing can stop our manifesting ability. If we do not have limiting beliefs about ourselves, we are free to realize our infinite Being and creative potential.

What Can We Know

If we can imagine living in perfection, we are already beyond space-time. Within space-time is where our ego-consciousness lives. Here there cannot be perfection, because perfection is part of infinity. In our imagination we can align with our inner knowing. This knowing is part of our essence, and it has no glitches. It can carry us into realization of our eternal presence of limitless awareness. Imagination that is aligned with ego-consciousness cannot realize this. To transcend space-time, we can go within to our inner knowing and meditate upon it. The truth arises here. Our inner knowing is beyond words and thought. It is realization. It is how we know what is real beyond the empirical matrix of duality.

When we ask for true realization, and we intend to become aware of our inner knowing, we can take steps to develop inner sensitivity. We can learn to pay attention to the feelings that arise immediately in a split-second in any encounter. These are prompts about how to face the encounter. Whatever we feel after that is an ego-reaction. Our inner knowing arises from the conscious life force flowing through our heart. It is our connection with universal consciousness.

Whether our current world is a creation of human consciousness, or is a computerized matrix, we are participating in it, and we can realize our consciousness existing beyond the empirical experience. This is the way to mastery of human life. We can have absolute alignment with our subconscious self in the energy of our heart. We gain the ability to regenerate our bodies and to direct the quality of every encounter without limitation.

If we cannot imagine perfection, we are logically limited, while in our essence we are unlimited. We have given control of our awareness to our ego-consciousness, so that we could fully participate in the human experience. When we feel that we want to know more, to know beyond our mental, emotional, physical and psychic limitations, we can take the inner journey. It is a path unique to each of us, and is the most exciting journey of our lives. Along the way we learn how to live in the moment, while being in compassion and love. We create joy in the lives of those we interact with.

As we begin to realize a higher dimension of living, our awareness expands beyond the empirical matrix. By learning the secret of the power of unconditional love, we gain access to the dominant creative power in existence, because of its positive frequency and amplitude. It can destabilize negative vibrations and correct defects. It maintains its vibratory level in every moment and cannot be influenced by negativity. We can experience joy, abundance and freedom when we align with our inner knowing.

Opening to the Expansiveness of Our Heart

To move beyond the restrictions, we have placed on ourselves, we can pay attention to the expansiveness of the energy of our heart. Unless we abuse it, our heart never runs out of energy. With as much enthusiasm as we need for great fulfillment, our heart constantly fills us with life force. Its loving quality of conscious life force flows to us as freely as we allow, within the limitations of our beliefs about ourselves and the toxins we have embodied. Our reception is based on the quality of energy that we have entertained and aligned with, while not realizing that our consciousness is vastly magnanimous.

There is great value to paying attention to the conscious energy of our heart. Once we recognize it, it sooths our anxieties and everything that irritates and angers us. Our heart vibrates with positive energy that is life-enhancing, and it exudes a conscious knowing that is more potent than our ego mind can imagine, yet it is subtle and unobtrusive in our awareness. It requires our clear attention in every moment, and it works through our feelings and our awareness. There is no thought, only knowing and realization. Thought then becomes the servant of our knowing.

When we live in the consciousness of our heart, we cannot be fooled by lies and other dissimulation. In our current embodiment, we are surrounded by falsehood and threats to our well-being. These are all negative energy patterns that we can recognize and transform in our personal lives, by maintaining our heart-centered focus. Because we are multidimensional beings, we can expect to live in a realm of miracles and wonders. Our ego-consciousness cannot imagine this and will discourage us. When we imagine that the threats to our well-being are overwhelming, our ego-consciousness will attempt to get us to align with the negative energy in order to resist it or to succumb to it with anger, depression, fatigue and loneliness.

Whatever our condition may be, the energy of our heart does

not change. It continues to provide us with conscious life-enhancing energy that can inspire us in every situation. It is our connection with divine consciousness and is our eternal guide in the form of our intuition. Unlimited by time and space, our intuitive knowing transcends the boundaries of our limiting beliefs. Everything becomes possible for us, because we are unlimited in our true Being. We only need to be willing to search within and to realize the personal truth of who we are. By constantly aligning in absolute confidence with the energy of unconditional love and joy, we can live in a higher energetic dimension and realize our expanding presence of awareness and mastery of life.

Exploring Our Innate Motivations

Our inner motivation arises from our innate being. Because our innate self matches its vibrations to our conscious awareness, we feel motivated to think about and do things that match the vibrations of our awareness. If we intentionally take our vibrations beyond our prior experience, our innate being adjusts the vibrations in our body and our environment to match our new way of being. There is a strong connection in our innate self to the energy of the heart of our Being. Because we are innately part of universal consciousness within the quantum field, our realization is being drawn to a higher realm, and we can invite our innate, our subconscious self to help us align with the consciousness of the Creator. It already is our consciousness. We live in the consciousness of the Creator, but we have been aware of only a compartment of it.

As we can begin to imagine experiences beyond duality, we must be willing to transcend our limiting beliefs about ourselves. Our consciousness envelops everything and is ever expanding. It is the consciousness of the Creator. Only our limitations keep us from realizing it. We need our limitations to be human in the realm of duality in order to continue to create it, but we do not have to limit our awareness to this realm. We are multidimensional, and innately we have unlimited awareness.

On the way to realizing our truth, we learn to experience more extreme energy patterns. As we regain our life force from supporting vibrational negativity, our experiences become more intense, and we enter a realm of greater love and joy. This path lies beyond personal drama and clinging onto our accustomed beliefs. If this is our path, we are being urged to be present in clear awareness, regardless of what happens. When we are aligned with our intuitive knowing, we are innately supported in every way. We just have to realize this and be in the moment. In every moment we can be aware of our larger situation, and accept everything that we personally experience with gratitude and compassion for ourselves. We are learning to expand our awareness into a realm of unconditional love.

Once we can realize what unconditional love really is, we can also know the extent of our consciousness and our creative abilities. There is nothing beyond our ability to recognize and make real in our experience. We hold the entire cosmos of universes and galaxies in our consciousness, right down to every subatomic entity. This is what we can open our realization to. And there is more. Beyond form and substance, there is infinite awareness. Beyond duality, there is a realm of love and joyous adventure. We are creating it by beginning to realize it in our imagination. Through our innate being, we share our energy in personal radiance with everyone we encounter. This form of enlightenment is living our lives from a higher-vibratory and positive perspective. Our motivations will naturally be in alignment with expanding awareness.

Aligning with Our Intuitive Knowing

We have realized that we have a hidden aspect of our consciousness. We call this our subconscious mind, but it is much more than a mind, and it is vast. It has access to every experience we've ever had in every lifetime everywhere in the cosmos. Through our intuition, we have access to all of this and much more. By recognizing the qualities of the energies that we intuitively sense, we can align ourselves with the feelings they stimulate in us. This prompts the subconscious aspect of our consciousness to resonate with us, and give us the bodies that we have and our experiences.

We have been limited by our beliefs about the limits of our conscious mind, but we are more than our conscious mind. We realize that our consciousness includes our subconscious, but it is not accessible for our ego-consciousness. These divisions of our conscious self are held by us as our limiting beliefs. In our true Self, we do not need limitations. Only our beliefs about ourselves keep them in place. We have made them real for ourselves in order to play the human show in all of its convincing scenarios, but we are more than this.

To begin to realize our true Being, we can choose to bring ourselves into the energy spectrum of the heart of our Being, the Source of our conscious life force. It is completely life-enhancing. It is the expression of unconditional love, which the ego-consciousness does not know. It is our intuition, which is also part of our consciousness, and it is our infinite presence of awareness.

Consciousness is universal. It is everywhere and in everything. We are part of it, and in our true Self, we have access to all of it, because we are the creators and modulators of energetic patterns. This is what we do. It is our nature. By our state of being, we are constantly creating in every moment by our alignment with energetic patterns through our attention and perspective. We are constantly contributing to the expansion of universal consciousness. We just haven't realized that we leave our mark everywhere and always.

We have allowed ourselves to be captured into a trance that encloses our awareness. We're still the creators, but weakly so, because we have been convinced to believe that the world of our human trance is all there is. Anything beyond is imaginary, and realizing its reality is beyond our belief.

By using our imagination, we can align with our intuition beyond the mental realm. It is the realm of knowing and is unlimited. It can carry our awareness beyond time and space, into the quantum field.

By shedding the beliefs in compartments in our awareness, we gain access to awareness of the expanse of consciousness and everything it contains and creates. We can know the awareness of all aware beings.

Opening to the Realization of a Higher Dimension

There is a level of energy that we are occupying, but that we are mostly unaware of in our ego-consciousness. In alignment with the energy of our heart, we can be aware of a dimension of loving and joyous encounters and creations. By feeling the quality of this energetic pattern, we can align our awareness with it and have a perspective of acceptance and compassionate understanding, while feeling the frequency of joy and deepest love. This energetic alignment can give us the realization of a higher-frequency realm beyond polarity and time/space, yet having control of time/space in our awareness.

We are all capable of transforming our experiences in our own reality. Everything that we believe is outside of ourselves, however we identify ourselves, is actually a projection of our own energetics, which we hold in our consciousness through our conscious and subconscious mental and emotional processes, utilizing the energetic spectrum of our beliefs. This is how we realize the vibrations of our personal reality, and this is what we can change intentionally, resulting in a different quality of reality for ourselves.

We can find that in the guidance that comes to us through our heart, we begin to know more and think less in our ego-consciousness. By living in emotional alignment with our intuition, we are free to create whatever we envision to enhance all aspects of our lives and the lives of all around us. When we can realize the reality of this state of Being, it fills our lives with its manifestations. Whenever we become aware of negative energy, we can accept it and transform it into light and release it from our awareness by refocusing on what we truly want in our deepest Being. This opens our awareness to a higher-vibratory dimension that we already occupy, but didn't recognize.

By releasing our fixation on the realm of duality, we can expand our awareness into a different dimension of energetics, one that is supportive of all of life and displays expressions of beauty and wonder. In our realization of who they are, we can become aware of the awareness of all beings, and we can interact from our heart with all who come into our awareness. Our inner radiance strengthens and brightens our aura, increasing our creative manifestations. We can play with all of the creatures and aspects of nature who encounter us.

All of the elements are within our domain. We can play with the water and the clouds and create their imaginary scenarios. We can align with the trees and other plants by opening our awareness to their awareness and knowing them intimately, while sharing their energies with ours. We can especially interact with crystals on their level of vibrations and know their awareness. There is so much that awaits our recognition and realization.

Living in High Vibrations

When we can imagine everything happening wonderfully in our lives in every moment, we transform into a higher-vibrating version of ourselves. And everything begins happening in fulfilling ways that we love and are grateful for. This is the beginning of mastery of human life. We transform from believing in limita-

Chapter 5. Understanding Our Expansive Self

tions for ourselves into Self-Realization in infinite awareness. We are so much greater than our physical presence, and to know this, we must open our realization to the energy of our heart. When we want to be aware of our inner knowing, it happens for us in great clarity through our emotional nature. We know when we're vibrating on the level of love and joy, as well as on the level of stress and doubt. We know all of these feelings so well and in the most subtle ways. By being aware of how we feel about ourselves in each moment, we know when we're vibrating and being stimulated in positive or negative ways. These feelings are always our choice for our creative energy.

We may feel that we're being influenced by our perceptions, but in reality, we're creating the quality of our perceptions in every moment. This is our nature. We are the modulators of energetic patterns, which manifest in our experience, based upon what we realize. When we learn to realize gratitude and joy in every moment, we can begin to live in a higher energetic dimension, unaffected by negativity. This all happens as a result of choices that we make about who we are. As we resolve and transcend our limiting beliefs about ourselves, we free ourselves from the constraints we have placed on our conscious awareness. Our limitations resolve into our eternal presence of awareness.

As we learn to be thankful for our true Self, our awareness grows into greater compassion and wisdom. Any negative energy that comes into our awareness requires our alignment for its existence for us. If we vibrate higher in positive ways, negativity cannot exist in our personal experience. When we realize this, we are no longer entranced in duality, and we become free to be the creators of a reality that is enhancing and invigorating for all of life. Our presence in the physical world becomes easy, because we know that we are always fulfilled in ways that we are thankful for.

Once we have resolved the self-imposed constraints upon our awareness, we are free to express the desires of our heart in limitless ways. We can become expansive in every way, with an infinite flow of powerfully radiant life force, expanding our etheric presence. We can feel this sense of personal radiance, and we know when we're experiencing these levels of vibrations. When we can intentionally vibrate at the level of gratitude and joy, our lives come into alignment with this energetic spectrum.

Inviting True Love into Our Experience

When we are in alignment with the vibrations of our heart, we are happy and content, and our lives flow effortlessly into experiences that we love. We can realize our true love, forming in the energy of the heart of our Being, gaining form in the etheric realm and manifesting in our experiences. By our realization of the reality of our true love, we fill our awareness with it, radiating its vibrations and attracting energetic patterns that resonate with us. This is the process of our creative nature operating in our experiences. We have the choice of intentionally imagining and feeling that we are living in gratitude, compassion, love and joy in every moment, regardless of our encounters.

Everything begins within us. We are the consciousness that creates everything in our lives. By organizing ourselves as members of humanity, we participate in the world that we hold in our realization, but we do not have to limit ourselves to the vibratory level of our species. Our consciousness is multidimensional, and our awareness can be free to roam many realms of reality, from great density to pure light. We have held ourselves in a prison of conscious limitations, but we have the opportunity to expand our awareness beyond the empirical world of duality. Nothing requires us to participate in the negativity of this realm. In order to participate in it, we need to create it. Without our fear, it cannot exist.

We have been held in duality by our fear of suffering and consciousness termination. These fears are self-created in our

Chapter 5. Understanding Our Expansive Self

imagining that, if we lose our ability to participate in the physical world, we have a lower quality of living, or perhaps even final termination. If we stop imagining this, and instead direct our attention to our eternal presence of awareness in our infinite creative essence, we open the prison that we have enslaved ourselves in. We can participate fully in the empirical world with personal positive polarity, leading to awareness of living in a realm that is completely responsive to our creative attention and alignment.

Although we have been largely unaware of it, there is a natural flow of energy enveloping us. If we can pay attention to it and feel ourselves living in its vibrations, we create experiences for ourselves that resonate with the natural flow of the conscious life force that we live in. We all know when we feel good, and we can use this awareness to imagine ourselves enjoying our lives with gratitude for everything that is fulfilling for us. Without fear, we free ourselves to love unconditionally, because we live beyond limitations. We can relate with the inner light of all whom we encounter, think about and feel. We transform the negative by bringing our light to it. By the power of our intentional focus, negativity either comes into alignment with our positivity or disappears from our personal experience.

6.

Living in Unlimited Awareness

When Our Lives Feel Easy and Right

When our lives unfold easily, and we're enjoying ourselves, it is because we have released our fears and are flowing in alignment with the energy of our essential Being. We are enhancing our experiences in alignment with Creator consciousness, as much as we allow. As we plan our lives, we can relax into knowing everything we need to know and do in every moment. We are the ones who determine the polarity and frequency of our vibrations in every situation, and we create the stress or ease of our experiences. When we are relaxed and at ease, even in otherwise stressful situations, we can be open and receptive to our intuitive knowing. When we allow ourselves to live completely intuitively, we create positive situations. We can have confidence that we always know how to interface with every situation, and how to create joy all around us, through the radiance of our state of being.

Stressful situations are creations of our ego-consciousness, and are based on negativity and fear on some level. Creations of the pure energy of the heart of our Being are always easy and joyous. By aligning with these vibrations, our lives become fulfilling in every way. When we realize that we are the creators of the qualities of energy all around and within us, we can realize how we feel about ourselves on a deep level. If we have stress in our life, it is because we are afraid of something.

Any fear is a result of our own imagining. Because fear is not present in Creator consciousness, we must create it for ourselves. The only way we could do this is to believe in our mortality, not knowing who we are in our essence. In order to realize that we are our presence of awareness beyond time and space, we must give up our attachments to personal limitations, by making intentional choices to do so. Then, in alignment with greater awareness, following up on resolving our fears whenever they arise.

When we no longer believe that we are mortal, because we have realized that we actually live multidimensionally, beyond duality and incarnation, we can truly enjoy being who we are, fully cared-for in every way, because we naturally create fulfillment all around and within. We are naturally brought opportunities and experiences that resonate with our state of being. When we are intentionally grateful, loving, compassionate and joyful, we attract experiences that elicit those feelings in us. As creators, it is our place to initiate the process by aligning with our heart energy and intuitive guidance.

Enhancing Our Consciousness

When we desire to expand our consciousness, we are naturally drawn toward unlimited awareness. It is our rising vibration that attracts more freedom and joy and enables us to recognize our limitations and how they are all negative in polarity and

Chapter 6. Living in Unlimited Awareness

based in fear. By facing them with a perspective of compassion and understanding, we can resolve them in kindness. We can be beings who are only positive, but who deeply understand the negative.

We can know whenever there is negative energy around. If we do not accept its polarity, it will disappear from our experience. It cannot exist for us, unless we realize it as real and give it our life force. When we do that, it immediately becomes real for us. We cannot realize its unreality, but we can maintain a positive perspective with gratitude and compassion for all and everything. Once we realize this, we can understand that the energetics are all for our enhancement in living.

There are many things that we can do to elevate our energy signature and our state of being, which is the source of our personal vibrations. Because we are infinite Being, we never have to stop expanding our awareness and having greater joy. What we do and how we live is irrelevant in this journey. It is our vibration in every moment that determines our destiny, and it constantly changes. The most important realization is our level of vibration in every moment.

If our goal is to align completely with our most enlivening and joyous vibrations, we must know our intuition. It always exists in connection with universal consciousness and guides us in unconditional love and infinite awareness. Because our intuition may be subtle, it requires openness and receptivity to higher consciousness. In every situation, intuitive guidance is immediate always. We're aware of it momentarily, and we can learn to focus on it, until we just naturally are constantly aware of it. We are front-running our ego and working with our innate being in acquiring a new skill. Gradually the ego stops attempting to control our consciousness. We can enhance this process with love and gratitude.

By paying attention to what we know within ourselves, we gain confidence that we are being sensitive to our intuition. This is important in being able to move beyond doubt. We may be guided to make ourselves over in many ways in order to express a higher aspect of ourselves. We're learning to play with energetics, which we control with our mental and emotional levels, our polarity and our confidence or doubt. Confidence is creative, and doubt is destructive to our creative energies. By learning to trust what we know, we can transform our lives in wonderful ways.

Learning to Be a Placid Lake at Sunset

We have learned to live in a world of significant limitations to our consciousness, but we can know our real identity and our great abilities. We can learn how to recognize them and use them. We can recognize how our empirical world becomes real for us, and what causes it to be. We are that cause.

We are such advanced Beings, that we have other consciousnesses working for us, paying attention to the vibratory level and polarity of our thoughts and feelings and manifesting these energies in our experiences. Our subconscious does not judge us. It translates our vibratory levels into the life forces of our bodies. Every defect that we harbor is a result of our alignment with the energetics of that defect. The energetics consist of how we feel about ourselves in every moment. These feelings reside deeply in our subconscious.

Making progress in life is complicated for us, because we keep disempowering ourselves with our doubt, shame, guilt, inferiority complexes, fear, anger and many other negative energy patterns. These are all imaginary aspects of ego-consciousness, creating our limited perspective on life and draining us of our life force by keeping us in states of anxiety, fear and other negative emotions. We have allowed ourselves to align with the resonant vibrations of life-diminishing limitations, and we provide our life force through our recognition of their reality for us.

Our limits are defined by our beliefs about ourselves. Regardless of how they arose, they each have a vibratory pattern containing a measure of fear and doubt. We can feel this and transform it into positive energy based on love by refocusing our attention, leaving the negative to fade out of our experience for lack of our alignment and life force. Without negative beliefs, we are left with only positive knowing. There is no more diminishment of our infinite Self.

As we resolve our limiting beliefs, our reality becomes infinite, as do our creative abilities. We begin to recognize that we live within an unlimited consciousness, enveloping everything that exists in all dimensions. We are constantly created in universal consciousness to express our own unique energy signature. As we encounter various levels of electromagnetic wave patterns in the unified quantum field, we have experiences in our awareness. The quality of these experiences is a result of our choice of our own vibratory level. As we gain a positive presence of awareness, we also gain the ability to hold a positive presence in any encounter. Our constant focus at the level of love transforms negative energy or dissolves it out of our presence.

Learning to Transform Our Lives

Our natural state of Being is everything that all of us truly want to experience. These should be the easiest experiences in our lives, yet we have resisted them. They represent the energy of our heart expressing itself through our mental and emotional abilities and interacting with other conscious beings. When we all can agree on what we most want to experience, we can know that these things are what is natural for us. We are created to experience them. We all feel good about living in love, joy, gratitude, abundance and freedom. We want these qualities in everyone and everything around us, as we experience the brightest, clearest, most fun, and enjoyable scenarios. Because all of these qualities are natural for us, we can choose to experience them in every moment by intentionally opening our awareness beyond our limiting beliefs.

Because we are all telepathic, and we want to live in our society, we have accepted the limiting beliefs of our community, but there is no requirement that we do so. It is all in our imagination and programmed into our subconscious. If we want to live in alignment with the energy of our heart, we can choose to resolve and transform our beliefs about ourselves, our identity and our nature. One way of doing this is to practice imagining the life-enhancing creator essence in everyone, including ourselves. That means living in unconditional love and gratitude as much as we can allow for ourselves, and to question everything that interferes with this.

By being the person that we truly want to be in every way, we can transform ourselves into expressions of our true Self incarnated with every cell filled with vitality and conscious life force directing it to be the best entity that it can be. All of our subatomic elements, atoms, molecules and cells are attracted to our energy and express themselves in alignment with our state of being. Everything and everyone in our awareness does the same thing. The way that we grow is by intentionally departing from our accustomed way of being and entering a new pattern of polarity and frequency.

In its essence, the energy of our heart expresses the consciousness of the Creator and is our guide for the enhancement of all life, including ours. By intentionally developing sensitivity to our intuitive knowing, we can successfully face every situation in every moment with gratitude and joy. Every experience exists for us when we recognize it and align with it in our attention and emotions. The energetic patterns existing in the quantum field manifest as experiences when we recognize them and realize their reality. By opening ourselves to their vibrations, we can intentionally align with them in our thoughts and feelings. We can imagine experiencing the kinds of energies that we like.

With practice we can realize that we actually are experiencing them, and they become real for us.

Living in Deepest Love

We are trained to have some level of fear at all times. This keeps us from experiencing true love, because fear keeps us negatively polarized, and because we doubt that it's possible for us to participate in universal, unconditional love. With fear and doubt we sabotage the fulfillment of our desires. They are all part of the consciousness of duality and can be transformed into positive polarity in our own intentional focus and alignment.

To transcend the complications of a world of duality, we can be open to recognizing positive energies and feeling their vibrations. We can choose to expand our awareness and alignment with a higher realm of good, even ecstatic, vibrations. To do this we can be sensitive to how every vibration in our awareness feels. We know immediately if the energy is positive or negative. The negative may seem to be powerful, but it has no more power than we give it. We create its reality in our consciousness with our recognition and belief in its empirical form. To make it real, we send it our creative life force through our resonance with it. We now know what it feels like to feel separated from our life-source and threatened with termination, but we no longer need to align with negative energy in order to be compassionately aware of it.

With the vibratory levels of our imagination and emotions, we are in charge of creating the qualities of our own experiences. We can create the forms of our experiences and as much else as we desire. By how we intentionally use our emotions, we automatically create how we are stimulated to feel. When we are positive and free of fear, we naturally manifest whatever experiences we desire, because we have only confidence. There is no other possibility in the positive dimension.

Because it operates in duality, the ego separates itself from higher consciousness. By directing the focus of our attention to positive, high vibrations, we can transcend the limitations of our ego-consciousness. Being in a state of acute, clear focus usually requires practice. Ultimately we can come into alignment with our natural state of Being in the consciousness of the Being who constantly creates us all in eternal, expanding awareness.

Once we have resolved our limiting beliefs about ourselves, we can be as expansive as we desire. In essence, we are our present awareness with the ability to direct our attention and vibratory alignment, resulting in our experiences in every dimension that we resonate with. To experience the greatest fulfillment, we can align ourselves in absolute confidence with the deepest love we can feel and know. We can live in a dimension of only positive, life-enhancing experiences. Unconditional love is inherent in our Being and is available for us to know and experience whenever we take the steps to open ourselves to it and intentionally align with its resonance.

Realizing Universal Consciousness

We exist within the consciousness of the Creator. In our greater Being, we can realize universal consciousness. We are individualized expressions, fractals, of the Creator, having free will to choose what to do with our awareness in every moment. Whatever we want, we can feel and imagine. Our consciousness is not bound by any constraints, except those that we place upon ourselves. We have created our ego consciousness to experience duality for our greater awareness, but we are not required to limit ourselves to this dimension of energetics.

Another dimension beyond duality awaits our recognition and participation. It is a realm that feels natural and good to us. There is only life-enhancing energy in everyone at this level of vibration. When we are able to realize this is our situation right now, we transform our lives. Everything changes moment-to-moment, as we focus our polarity and vibratory resonance.

We are the One. We are the consciousness of the ocean. We are the consciousness of the Spirit of the Earth. We are the consciousness of all living beings. We can learn to feel how they feel. All of this comes through our intuition, when we seek it and are open to knowing it. It works through our imagination and emotions.

We can feel this connection through the conscious life force that flows to us through our heart. We call it unconditional love, and it is our true unlimited awareness. Our awareness also holds our creative ability, which we constantly express in our state of being. It is essentially how we feel about ourselves in every moment. This is always our personal choice. If we have difficulty with this, we can carefully examine our limiting belief and resolve it through compassion and intuitive knowing.

If, for any reason, we feel negative about ourselves, we can change this at any time by intentionally choosing to be positive and to enter the state of gratitude and life-enhancing energies. No one requires us to feel negative about ourselves, regardless of what kind of energies we have aligned ourselves with. When we decide to open ourselves to the vibrations of our heart and to feel our essence within the consciousness of infinite Being, we can begin to be masters of our life in any dimension.

What Can Happen without Limitations

If we are interested in resolving our limitations, so that we can be completely free in every way, it is possible by intentionally being positive in every way. It is training our perceptions to be aware of positive energy everywhere. It is all present in our consciousness, awaiting our awareness and expectation in complete confidence. To make this expansion possible, we need to adjust our polarity to completely positive.

The ego consciousness is not helpful with this change, because the ego feels vulnerable. All of its limitations prevent it from being mostly positive. It cannot exist without fear. This change in polarity is a radical change in our perspective and needs our intuitive guidance. We can train ourselves to be able to control and call forth our emotions and thoughts and just be present awareness. This is what mastery requires.

Being positive entails feeling confident and at ease in every moment. Any threats from negative entities have no effect, because we're no longer attracted to negative polarity, and they cannot become positive. We are beyond their dimensional veil. This is the reason that the jump to completely positive is such a challenge. The ego does not want to let go of control, even though it has no real choice, because we are the masters of our ego, which has no higher consciousness.

As we resolve our limiting beliefs about ourselves, our ego dissolves, because it consists of our limiting beliefs. Without limiting beliefs, we are able to create everything we need, and everything our heart desires, because we hold no negative energy to defeat our creativity. We become free to express the energy of our heart and thoroughly enjoy being human in a completely loving and joyful way.

A Pathway to Greater Mastery

If we desire to become aware of other realms and dimensional levels, as well as our own deeper consciousness, we must learn to meditate deeply. At first this requires concentration in developing awareness of the inner path. We can deepen our sensitivity to our intuitive knowing and greater extra-sensory abilities. By breathing deeply and rhythmically a few times with the intention of expanding into our infinite awareness and imagining ourselves enveloping the cosmos, we can feel ourselves completely at ease just being present awareness. With practice we can achieve this state of being in which we are unfocused and receptively aware. Our awareness can expand as far as we are willing to go. We are our only limitations. It is important to be aware of our intuitive guidance and to stay positive and open to high vibratory feelings.

Whatever limiting beliefs are hiding in our subconscious will reveal themselves as we begin to transcend them. We can face them with open hearts and gratitude for their intentions to protect our ego-selves. Once we recognize them, we can understand that they have been necessary for us to have an authentic human experience in a realm of duality, and they cling to our psyche as parasitic attachments. They need our life force through our alignment with them in order to exist. When we become completely positive, we no longer align with their negativity, we reclaim our life force, and they disappear from our experience, opening our awareness beyond our former limitations.

We may have to work on this process for a while before noticeable progress, but as we become sensitive to our intuition, we gain deeper understanding of everything. We can feel more radiant, and we love to look into each other's eyes, where we exchange conscious photons. As we learn to be at ease and aware of our intuitive guidance, a more expansive life opens up for us. This requires a strong intent and lots of practice, and the result is ascension into a higher realm that is only positive.

We can learn to control our thoughts and emotions positively at will in alignment with our intuition, creating and receiving everything we need in every moment. We can communicate intuitively with our guides and higher beings, and we can thoroughly enjoy each moment and be masters of our every encounter with others. We can realize that we are our eternal, present awareness, able to manifest however we desire anywhere, while also participating as radiant Beings in our human experiences.

Living in Expanded Consciousness

We can live in compassion and joy. This positive level of vibrations draws us into our intuitive knowing of expanded awareness of a higher dimension of living. We can be in any emotional and mental state that we choose. It is this state of being that creates the vibrations of our experiences. We can transform our lives by intending always to be positive and understanding. This intention opens our receptivity to the energetic patterns that are available for our recognition. Our alignment with this level of energetics opens up for us a loving way of life with others who are loving in their lives.

Everything arises from consciousness. We are designed to be able to modulate into experiences the energetic patterns that come into our awareness. Keeping our awareness within the empirical compartment of consciousness creates a longer stay there. By changing polarity to only positive we can move beyond time and space. This is where our creative power begins in our imagination. Everything we experience is filtered through our imagination, creating the kind of experiences we imagine as real. By imagining as real something we haven't yet experienced, we create the experience on the same vibratory level. Being able to confidently feel and know that we are in the experience, is how we create the quality of the experience that manifests in our lives.

We do not need random thoughts filling our minds. We can take control of our consciousness by learning to control our emotions and imagination. We can practice and become clear in present awareness. Listening to the inner sound current is helpful in being present and aware of the energetics of our own consciousness. Developing sensitivity through open receptivity to and expectancy for greater love opens us to greater insight into every moment.

While moving into higher vibrations of love and joy, we fade into a higher dimension of energetics in a parallel timeline. Each of us can make this shift at any time by aligning completely with our intuitive guidance, because this is where our conscious life force is taking us. We can still choose the path of living in duality, if only to visit, once we know how to expand back into our eternal, infinite awareness.

We are on the path to expansion with higher guidance. In aligning with our intuition, we transcend the ego and all of our limitations. As they arise, we can be aware of the quality of their energy and decide if we want to continue to give them existence. Because we recognize them as real, we give them our life force. There's no requirement that we do so. It's a matter of choice on our part, if we want limitations, or if we want expansion.

Whatever is not eternal here is a human creation and can be changed in our personal experience by the vibrations of our state of being. We can practice just being present awareness and listening to the inner sound current. We can practice calling up emotions and creating visions of our new world of expanded consciousness in joyful ecstasy of beauty and love, until we realize that it's real, and we're experiencing it.

Penetrating Deeper into Our Consciousness

We express ourselves as energetic patterns that attract our experiences. It all happens within our own consciousness. Our energetic alignment with the world that we know and experience has kept us focused on the realm of duality. The realm of only positive energetics is present for us, when we can recognize it as real and intentionally align with it. In this level of conscious awareness, we can feel the energy of the heart of our Being. It is life-enhancing and regenerating in vibrant ways. These energies are not limited to time and space. They are universal, as in the song of One.

Aligning with heart-energy has its own inner sound current, which changes with our level of consciousness, as well as feelings of gratitude and joy. It opens our intuitive sensitivity and awareness and keeps us present in the moment. We can become clear and passionate. We can learn to call up emotional states in subtle and powerful ways, accompanied by imaginary scenarios. All of our experiences actually are imaginary, and we experience the ones that we believe are real.

While we're embodied on the Earth, we live within a limited spectrum of our potential awareness. These limits exist only within our own consciousness. We can keep them, if we want to, or we can transcend them to explore our infinite Being. If we are open to it, this can be accomplished by choosing to be always positive, loving and compassionate. By focusing on these energetic expressions, we can go deeper emotionally and with greater understanding. Aligning with the energy of our true heart allows us to recognize the light in everyone and to interact on this level.

The magnetic poles of the Earth are changing as the negative is receding. Our planet is becoming unipolar at the center. This change in energetics is affecting us as well. We are being urged to be only positive. In alignment with the Earth, we can be moving toward a deep understanding and feeling of unconditional love without limits. This is the energy of our heart, which also offers us constant guidance in our intuition, whenever we seek it and are receptive to it.

As we become more positive, we are able to enter a world of love and peace, because these are the energetics that we express and attract. We enter this world by recognizing its reality and aligning with its vibrations, causing it to manifest in our experience. We can do this while living our accustomed lives and making adjustments as we go along, or we can just make the leap in consciousness and quickly transform our lives, possibly with some challenging pushback that we can resolve with intuitive guidance.

Choosing Negativity or Positivity in Each Moment

When developing inner sensitivity to our intuition, the first thing that we hear is our inner sound current. That lets us know that we're in the right vibratory resonance. As long as we can align with the energy of our heart, we can be aware of this sound, which disappears when we change to negative in our momentary energetic alignment. When we ask for guidance within, we need to pay attention, because intuitive guidance is instantaneous or even before we complete our request. It comes most strongly through our feelings and requires openness and clarity for true understanding.

We can begin this process by being aware of how we feel about every decision we make, and every decision that someone else makes for us, and that we acquiesce to. These decisions represent our ego's attempts to achieve fulfillment, while also fearing termination. Fulfillment is a positive goal and cannot be achieved by a negative orientation. In order to reach fulfillment, we can be only positive. We are essentially beings of love and compassion.

If we keep searching for how we would like to feel in every encounter, regardless of what kind of energy we're facing, we can imagine feeling positive and experiencing positive energy, even if we're encountering negative energy. Once we begin walking the inner path, every encounter becomes guidance for being sensitive to our intuition and following it. It is the awakening of the unique genius for each of us.

When we resolve our limiting beliefs about ourselves, and we imagine expanding our awareness to our unlimited Self, we can train ourselves to become aware of a higher dimension of living. When we recognize it as real, this becomes our experience. It comes from being only positive and requires practice and confidence that becomes knowing. A transformation of our psyche occurs.

As we expand into our true Self, we can confidently know

that we are infinitely powerful creators. This enables us to enhance the lives of everyone we encounter with the radiance of unconditional love and deep understanding. In this way we can transform humanity by repolarizing and raising the energetic signature of our species.

Achieving Mastery of Ourselves

Although we may not realize it, we are the masters of our lives in every way. As humans, we are subject to a trick that we play upon our limited, personal consciousness, and so we don't believe that our conscious presence of awareness is unlimited. We have limited our awareness by believing that we are limited, and we have lived our lives with this perspective, not even able to imagine being unlimited.

We are designed to be guided by our intuition, which is unlimited in wisdom and understanding. When we have aligned ourselves with our intuitive knowing, our hearts can open in true love and joy. We can realize ourselves in the eternal Consciousness that envelops all. Our awareness in universal consciousness transcends all limitations, dissolving them in our human consciousness, and allowing us to direct our lives in conscious mastery.

Aligning ourselves with our intuitive knowing is our path to mastery. This can happen if we direct our attention to finding the positive vibrations of love, gratitude and life-enhancing feelings. When we can bring these feelings up within ourselves, we can be aware of our inner knowing. This is the natural vibratory level of our expanded being, beyond polarity, time and space.

We are designed to be creators and experiencers of the energy of our creations. We create with our conscious awareness through the energetic patterns that we align with in resonance. When we are holding any level of fear, we create experiences that stimulate fear. When we are completely positive, we create only situations that elicit joy and appreciation. We do this through our state of being, the qualities of our thoughts and emotions.

Our intuition is only positive, and its guidance is life-enhancing. Since it is beyond the capabilities of our ego, we may have inner conflict. We can learn to pay attention to our immediate knowing, regardless of what the ego may want. The more we recognize our intuitive guidance, the greater our awareness can become, until we can realize our presence beyond time and space.

Expanding Our Consciousness

We initially chose to have the Earth human experience in order to know how the vibrations feel in this spectrum of energetics. We became entranced and developed beliefs that restrict our awareness to living in duality, allowing us to experience all kinds of fear. With our limiting beliefs, we formed our egos, and they keep us from being able to realize our true, expansive Self. If we want to resolve this situation, we can develop sensitivity to our heart's intuition. It constantly guides us to feel ourselves being present in a field of support for all aspects of our lives. These energies are directed to create the kinds of experiences that we recognize and believe are real. This is purely a psychological process that manifests as physical.

When we open ourselves to the positive energy of our heart, we can align with our intuitive knowing. Its guidance is always life-enhancing in the direction we desire to go in our deepest Self. It is the conscious presence of the Creator, providing our conscious vitality for us to use as we choose.

While we explore and experience different levels of vibrations in our encounters and within, we are living a realm of duality, predominantly based in fear.

In the perspective based in fear, we are always in need of something, because we believe our sustenance is outside of ourselves. In feeling limited, we have been receptive only to some of the life force we receive through the heart of our Being. By aligning with the energetic level of love and gratitude, we can open ourselves to awareness of our eternal, unlimited presence. This awareness is being supported by the shift in the Spirit of Earth to positive with rising resonant frequencies. We can best feel these when we are alone in nature, especially in beautiful and majestic places.

The ego uses our life force to keep us imprisoned within the compartmentalized consciousness of the human world. Staying within these limits of awareness is our choice, and we can be adventurers in consciousness. A good place to begin is with deep, rhythmic breathing and then sitting comfortably while practicing just being present awareness. We can explore different meditative techniques, until we expand beyond time and space.

We can choose to be only positive and compassionate always, even in the face of strong negativity, and we transform our lives completely. All aspects of our lives become rich and vibrant, manifested as a result of the vibratory resonance of our state of being. We do this be staying positive always, with deep understanding through the higher guidance that we come to know in the presence of the Master within ourselves.

Being at Peace and Living with the Results

We can all come to a place of peace within. This leads to peace with everyone else and our entire environment. As we do this, we elevate our personal vibrations, which affects everyone around us. When we have no negative energy, we are radiant with love and compassion. We even become transformative influences in the presence of predators and parasites. If

Chapter 6. Living in Unlimited Awareness

we can maintain our vibratory level, negative energies cannot exist in our presence. As we open to the unconditional love of the creative Source of our Being, we embody the life force that enhances everyone.

We can express the true passion of our heart in all aspects of our lives. We can naturally accept, compassionately forgive and powerfully radiate unconditional love, while feeling a deep connection with everyone we encounter, including those with evil intent. They are the ones who have separated themselves from the Source of their life and now need our life force. This they cannot have, if we maintain our divine radiance, for the energetics are oppositely polarized and out of resonance.

The only way for us to be subject to threats and intimidation is for us to accede to their level of energetics by aligning ourselves with them, in order to join or oppose them. In either case, we share our life force with them through our attention and negative alignment. With our positive orientation and our intuitive alignment, the negative has no place in our consciousness and no access to our life force.

When we are completely positive and conscious of our eternal, present awareness, we can become powerfully passionate, because we are completely confident of everything we think, feel and do. We can feel the unconditional love enveloping us in the quantum field of all potentialities. Our hearts are pure, and we always feel wonderful. We have no personal needs, because we create what we need whenever we think and desire it. We deeply know that we are powerful creators just by our constant positive vibrations, which we identify with in our thoughts, feelings and perspective in every moment.

This is the quality of life that we are being directed toward by the increasingly only positive polarity of Gaia, her increasing vibratory resonance, and the great intensity of incoming gamma-ray conscious photons. Our Sun is growing in intensity and brightness, as is our entire cosmic environment. We are being carried into higher consciousness in every area of life. Currently, all of this is subtle, but perceptible, especially in our emotional awareness. Once we have resolved our self-limiting beliefs, this awareness can become acute.

Guided by our all-knowing intuition, we can stay in positive, high-vibrational conscious awareness and always know everything we need in any focus of attention. We can understand that human life as we currently experience it is just a game that we're playing together in our consciousness. We can have fun with it in so many ways, whatever we enjoy. By living with the intent to uplift our feelings and be inspired, we radiate this energy and experience its resonance in our encounters with others and in the love and fulfillment that comes into our lives.

Living in Our Mind and in Our Heart

Our human mind operates through our brain, which has a neural network designed to function in the world of duality and separation. We have two hemispheres that understand left and right, sinister and forthright, bad and good, desirable and undesirable, positive and negative, self and other. We have created mental languages that emphasize the polarities and differences in our experiences. We have used our mind as a tool of our ego-consciousness, but it has a greater essence.

Our intuition operates through our physical heart, which has a neural network that is much more powerful than that of the brain, and it functions in unipolarity of positive only. It is non-judgmental and knows only goodness and unconditional love. It is intimately connected with our subconscious innate being. Our heart enlivens all parts of our being, physical, mental, emotional and etheric, with the life force flowing from the consciousness of the Creator of all.

Our heart is a direct expression of the essence of our greater Self, of whom we have been largely unaware, due to the needs of experiencing human separation and duality. When we are ready and desirous of expanding our awareness into greater joy and abundant living, we can resolve our limiting beliefs through our intuitive knowing, and we can open ourselves to awareness beyond time and space. Once we have become aware of our eternal presence, we can choose to continue to participate in our human experience, and we can choose to have fewer limitations whenever we want.

The ability of living in creative mastery of the human experience is available to us now. It's a conscious intention to be always positive with no mental and emotional encumbrances. This gives us a state of being in clarity, resulting in great sensitivity to intuitive guidance. The guidance of the heart of our true Being is always given in deepest love and enrichment in every way. We can trust it absolutely, and as a result, we can trust ourselves to be true always. We can have all of our needs and desires fulfilled by our own creative spirit, and we can realize that we are part of the consciousness that creates and enlivens everyone and everything. We are all the same Being with infinite creative power.

This gives us a transformed perspective on our role in the play that we're acting in, designing and producing. It's all within our own consciousness and within our own energetic choosing.

Enhancing Our Human Experience

From Quantum physics we know that there is a creative, universal consciousness that is the source and essence of everything that we know of. Everything is conscious, even the most minute sub-atomic waves/particles, all atoms and molecules, even rocks. We are surrounded by conscious beings appearing as empirical entities, many of which we have falsely regarded as lifeless, including our planet and all of its constituents. There is much we may surmise about universal consciousness and the Being that it emanates from.

The Being that has universal consciousness creates other

Beings within Its own consciousness, and all of these Beings share in Its consciousness. Some are designed to expand the Creator's consciousness with unique experiences and vibratory patterns. These Beings have unlimited free choice in everything they experience. They can think and feel any vibratory patterns that they are interested in or entranced by. They can imagine new experiences with interesting feelings. They create experiences by their presence of awareness, creating emotional alignments with energetic patterns. Through their attention, they share their life force with the energetic patterns that vibrate in resonance, stimulating these Beings to recognize these energies and bring them into physical expression. In this way, they can create anything they desire and feel.

These Beings are us, when we shed our attachments to the limitations we believe we have. The creative essence of universal consciousness is life-enhancing and materializes for us the qualities of our state of being. When we live in gratitude for everyone and everything in our experience, and in everything we imagine, we vibrate in a positive way in resonance with the unconditional love of the enhancement of all conscious life. We experience everything we could need and want. Our abilities allow for us to live in as much abundance and mastery in our lives as we allow ourselves and believe are real. We have created and maintained all of our personal limitations by our imaginary beliefs, which we accepted in order to participate in the dualistic human experience.

Our human lives are one of many expressions of our individual consciousness. We are multidimensional and express ourselves beyond space and time as other persons in as many timelines as we desire. In this human life, we are provided with manifestations of the vibratory levels that we pay attention to. By living in gratitude and knowing intuitively that we are whoever we imagine and believe ourselves to be, we can train ourselves to transcend our limitations, until they become unbelievable. With no attachments, we can be clear and receptive to our inner knowing, which guides us to mastery in all aspects of our lives.

Feeling the Warmth of Our Guidance

We can know and feel our inner divine guidance in every moment. Our eternal Self radiates its presence in our heart energy. When we are open and intentionally receptive, we are drawn into positive, higher vibrations in our thoughts and emotions. We have control over our attention, and when we focus on high-vibration scenarios, we attract compatible energy into our experience. In our human lives, we have developed a perspective of appreciation for the depth of darkness that can be suffered, and for our feeling of aversion toward it. We know what we want, because there's a feeling of warmth toward it. It draws us into imagining something heart-felt and feeling ourselves experiencing it. This is how our inner guidance helps us to create what we want.

These feelings are natural for us. We don't have to do anything to notice them, except to be aware. As we go through our lives, everything signifies something for us. If we are aware of this, we can know where our attention brings the greatest warmth for us and directs us to toward our true destiny. By holding this perspective with gratitude, we elevate our experiences, as we journey toward fulfilling our lives in every way.

Being sensitive to our intuitive guidance means paying attention to what we know deep within, while participating in the show all around us. We can hear our inner sound current, which stabilizes us emotionally, while allowing us to feel the warmth of our guidance. The energy that comes to us through our heart is completely life-enhancing. It arises from universal consciousness and conjoins us with the conscious awareness of all beings in the unity of spirit in unconditional love.

As we open to this experience, it can be a bit dizzying, because all the limits that we have lived within are gone. We're navigating our lives essentially based on how and what we feel and what we observe and imagine, with no worries or fears or doubts about how we're doing. We can be aware of the signs to guide us, and sometimes even verbal directions, when needed.

Living by intuitive guidance beyond ego-consciousness can carry us beyond our limited personalities and body consciousness and into awareness of our expanded Being in full Self-Realization. In our constant present awareness, open to intuitive knowing and feeling, we can realize our infinite creative ability. When we are moved to create something, we can realize what it is and feel ourselves experiencing it. Then we let it incubate, as our energy signature radiates the energy into the quantum field for manifestation, and we go about our lives and onto new experiences and challenges.

Leaping into the Unknown

When we decide to be our Higher Self, we could not know all of the implications. We just know that we want to be joyful and loving and deeply loved. We want to feel really good, even ecstatic, for as long as we desire. We want everyone in our awareness to be in alignment with these energies. To be able to be at that energetic level and state of being, we find that we have to let go of our attachments to our personal limitations. These are rooted deeply in our consciousness, and they need to be loved for the experiences they have given us in deepening our compassion.

The rising energetics of our planet and our cosmic environment are drawing us into more positive and loving relationships and feelings about ourselves. We can begin to realize that our awareness does not need to be limited to our ego-consciousness and even to the empirical world. These limitations are of our own making, and we can release them when we desire. If we keep imagining ourselves as unlimited as we can, immersed in gratitude and joy, apart from any concerns, passions or limited expectations, we will be drawn into experiences that resonate with our expanding vibratory level.

Every life form here is created to resonate with the vibrations of the Spirit of the Earth. Everyone is sustained and guided by the higher consciousness of its species, unless humans interfere with it. No one is created to deal with catastrophe, but we have chosen to create and participate in this energy in order to deepen our personal experiences of fear in developing greater compassion and understanding. Humanity also has a higher consciousness of our species. We have a group consciousness that is becoming more positive and expansive.

Many spiritual masters have expanded the consciousness of humanity into a dimension of unconditional love and infinite creative ability. We are not the first to enter these levels of vibration. They already exist and are held in the awareness of the Masters. They are available for our realization. We can develop our own vibratory level by intentionally imagining and being the One we truly want to be. We can be drawn to be unlimited in awareness beyond time and space as pure, infinite, present awareness.

Translating awareness of infinite Being into our human self comes with higher guidance in our inner knowing. Our true intuition is ever-present and all-knowing, when we know we are in alignment with the energetic level of the spiritual Masters. This is also our connection with the consciousness that creates everything. We can be intentionally attracted to our true Self-Realization.

Transforming from Victimhood to Mastery of Life

Human society has programmed us to recognize ourselves as victims of our circumstances, especially in the face of powerful entities beyond ourselves, yet this is true only if we accept it. It is possible for us to be absolutely sovereign in our personal life and to radiate personal freedom into the energy signature of humanity. Once we learn the truth about ourselves, we can initiate the process of actualizing it in our experience. If we want to live in personal freedom, we must change our beliefs about ourselves in order to transform our lives.

We experience only the energy within our own consciousness. Because consciousness is universal, we participate in it only as much as we can recognize and believe is real. This is how we limit our awareness and create our experiences. Expanding our awareness can be done through intentional practices of deep, rhythmic breathing, meditation, bio-feedback, mind-altering substances and out-of-body experiences, such as dying and returning. Other methods can include sudden flashes of insight and enlightenment and just having a determined intention of opening our awareness and following intuitive guidance. However we decide to awaken to our true potential, it can happen with our motivation to retrain ourselves, resolve our limiting beliefs, open ourselves to infinite awareness and follow our intuitive guidance.

Ego-consciousness is centered in the conscious mind. It functions in the empirical world and uses logic. It thinks by aligning with the energetics of thought patterns in the quantum field. It doesn't matter whose thoughts we are thinking. It is our energetic alignment that is important. This is how we attune to thoughts and ideas and provide the basis for making decisions.

As we become more aware of our intuition, we can realize that it is present in our physical senses as well as our thoughts and emotions. It is our deepest and most comprehensive knowing. Often it is symbolic in its portrayals. What we get from our intuition is a function of our openness, clarity, visionary ability and emotional control. If we practice these states of being and abilities, all of them are possible for us.

Beyond time and space, our expanded conscious awareness is infinite, but the ego cannot believe or even imagine that this is true. As a result, our human conscious awareness is limited to time and space. This limitation is within our control. To resolve it, we can intentionally imagine the best life experiences and relationships that we would like. As we become more positive and joyful, we can continue to expand our awareness toward infinity in every possible aspect of experience. Once we have resolved our limitations, we are no longer victims. Guided by our inner knowing and feeling, we become the masters of our experiences.

Our Destiny of Vitality and Mastery

We are destined to be filled with vitality and love, even though our experience on Earth has been fearful and disillusioning. We've taken this excursion into duality in order to experience the feelings of separation from our Source of Life and to know all aspects of destructive energies. As a result, we have been able to develop great compassion, mercy, wisdom and fortitude. Only now are we learning that it is all an illusion that we have allowed ourselves to believe is real.

The entire empirical world is a play in our consciousness. There is nothing solid about it. It consists of moving patterns of electromagnetic waves that we perceive as sensory stimulation. Although we have been unaware of this, all of our interactions are by choice on our part, even those that threaten us; however, our conscious presence of awareness cannot be threatened. Our reality is beyond time and space, and we can learn how to master our physical world and all of our interactions.

Because we are fractals of the Creator, we can change the energies around and within us. We can choose in every moment how we feel and what we think. Thoughts and feelings are part of our essential being, and they are completely within our command. If we choose to be only positive, we open ourselves to thoughts and emotions that are only positive. We take ourselves out of the realm of shame, anger, fear and doubt, and we enter the dimension of love. Here we are only love in the truest sense.

By imagining and feeling ourselves expanding in love and joy, and asking our guides and angels and our higher Self to draw us into the presence of the divine, we can transform the energies that we encounter. We just have to be true to our higher nature and follow our intuitive guidance, which we know through the energy of our heart. In every aspect of our lives, we can intend to enhance the life of everyone in our awareness. We can intend to interact only with the divine Being in everyone.

By inviting the feelings of abundant life and joy into our awareness, we can release all negativity and limiting beliefs about ourselves. When we move beyond polarity, we cannot be both dark and light. If we are light beings, we are beyond fear and duality, and life must arrange itself to align with our vibratory level. When we are able to anchor our awareness in positivity and love, and hold our alignment with this vibratory level, we begin to expand into higher consciousness and are able to direct our infinite creative ability.

Processing Our Spiritual Growth

In the face of attacks on many levels from governments around the world upon their people, we can surmise that a showdown is coming. In contrast, our entire galaxy is becoming brighter and more positive. Our enveloping energies are drawing us toward more love and compassion. The light is strengthening, and the dark is moving into another, more-negative dimension. As everything that is negative and dark is being brought into the light, negativity is losing its place in our galaxy. Because much is coming into the light, the observer may think that everything is getting worse, but the natural energies are flowing toward more positive vibrations, and we can make great progress in that direction.

By being higher versions of ourselves, we can be elevating the energies of humanity and making everything more difficult for negativity. By withdrawing our life force from the psychopaths, we are opening to the light within and around us. As we become receptive to the light of our intuition, we are guided wisely in allowing the negative to disappear from our lives. Our natural tendency now is to become enlightened. Everyone who is repairable will drift this way. Along the way we'll have some jolts that will quicken the process of becoming completely positive.

Being only positive is clarifying us, so that we can be aware of our deepest, darkest secrets and fears. Then we can resolve them with acceptance, forgiveness for ourselves, gratitude and love for the experience. We can release all of the energies that we created while learning how to be and live. We didn't know there is more to living than the experience itself. We didn't know that we're unlimited and are free to create any energy patterns that we want with our mental and emotional abilities.

Once we gather our abilities and begin to realize our true Self, in our infinite awareness, we can realize our invincibility in eternal being. Fear has none of our life force and disappears. By following our intuitive knowing, we can become completely love-centered and filled with gratitude, joy and compassion. We become able to trust ourselves implicitly, freeing our creative ability to manifest our heart's desires. We can live in a dimension in which negativity cannot exist, and we cannot be threatened. We can experience only goodness and mercy always in the consciousness of the Creator.

Living in the Consciousness of the Creator

As we shift our awareness to realizing our presence in the consciousness of the Creator, we can recognize the need for clarity in wielding our creative power. If we allow ourselves to waiver into negativity, we immediately create imperfection in our energetics. We cannot be both positive and negative at the same time.

That is duality, which doesn't exist, except in our own creation. We've been creating it for eons, but we can change that to only positive at any time by making the choice.

If we can release our desire for something negative and limiting, we can free ourselves to align with the energies of unconditional love, our natural state of Being, because it is part of divine consciousness. We can know our deep connection with every conscious presence, which is everything that exists. We are all living in the consciousness of the Creator. When we can know the reality of this, we are fulfilled in every way, so that we have no cravings or needs. By aligning with our intuition, we are guided to everything that we would want. Nothing belongs to us, but all is given to us for our use and enjoyment in the moment of our alignment. When we no longer need it for our interactions, it disappears from our experience.

By learning to pay attention to high-vibratory energy patterns at the level of love, gratitude, compassion and joy, we can live in a state of ecstasy and bliss as much as we choose. This opens us to increasingly powerful vitality, because we are aware of greater wonders and more beauty in music and sensations. Our abilities are intensified, and our energetic radiation expands. As we become more open and energetic, more wondrous experiences come to us.

Once we know that we can absolutely trust ourselves to be true in every moment, we become invincible, living in the clarity of divinely-guided intuition. With our intentional direction, we can transform any energetic pattern into alignment with us or send it into the Light. We can come into a deeper understanding of life, aware of our Source and our connections with all conscious beings. Expanding throughout universal consciousness, we become aware of the awareness of others, imbodied or just energetic.

Our process of making these adjustments in our lives may have been slow, because of the density of our world. We're accustomed to being compromised in some way, not letting go of attachments to limitations, even though they may not be satisfying. None of them are real, except for our beliefs, and they're deeply set in our consciousness. By paying attention to our intuition, we can resolve them and transcend their vibrations into our expanded Self.

Finding Our Inner Truth

What we recognize as the reality of human life is actually a play in consciousness that we have all agreed to participate in. We have the ability to change our entire situation at any time. Although we have a script that we follow and certain experiences that we have agreed to participate in for the enhancement of our character, we have great freedom of choice in what transpires in our own being and how this influences our experiences. We are created to magnify the experience of our Creator, Who lives through us in the complete spectrum of our energies.

Although in our lives, as we have known them, we have been unaware of the divine presence within us, we can awaken to our true essence, whenever we are ready. The entire drama that unfolds around us is only a distraction that keeps us occupied, so that we cannot realize who we really are. We have learned to subject ourselves to believing that we are separate individuals apart from the Source of our life. If we seriously think about this situation, it becomes obvious that it is impossible. How can we be alive and be separate from the source of our life?

Our challenge is to find the divine presence within ourselves. If we search for the part of us that is the most life-enhancing, we come to the energy symbolized and embodied in our heart, which lives entirely to benefit us, regardless of what we do to it. It is the most powerful organ in our body by far, and it has its own consciousness that we can be aware of and align with. Our heart has a greater consciousness than our brain, which is the seat of our ego-consciousness. Our heart is filled with the presence of the Creator, and it constantly offers guidance in a non-intrusive way for our fulfillment and greatest love.

For the greatest experience in duality, we have been entirely entranced by our ego-self, which is aggressively eager for control over our life. We have not recognized the guidance of our heart, which we know as our intuition. Our intuition is unlimited in its consciousness, and we can align with it by realizing its positive polarity and high vibratory level. It is completely positive and life-enhancing in every way. Unless we align with its energy, we cannot be aware of it. We must be open and receptive to love in the highest and most refined way and ready to be true to its guidance.

By realizing the presence of the divine within us, we can transform our lives. No longer are we subject to karma and negativity. When we are stuck in the limitations of ego-consciousness, we cannot experience these higher energies and do not even recognize that they are possible. We can open ourselves to the wonders of great love and joy and the presence of miracles, and we can realize the great Being of Light within ourselves.

The Energetics of Shifting Dimensions

We are constantly created in unconditional love and given our conscious life force and our unique Being in every moment for all eternity. In every present moment we have our unique conscious awareness, both in the body and beyond time and space. There is an expansion of human awareness beyond the empirical world. In our inner knowing, we can realize expanding into greater awareness of a brighter life of gratitude, beauty and joy.

The consciousness of the Creator is drawing us toward positive polarity in love, freedom and abundance. For us to realize this and experience it in our lives, we can open ourselves to the expanding flow of life and confidently intend to participate in it now. It already exists for us as patterns of energy, but we must recognize and realize it, in order for it to be real for us. This may sound like a trick of consciousness, but it's actually penetrating the limitations we have placed on our conscious awareness.

If we want to experience the enhancement of life, we can align with the energy radiating from the heart of our Being and coming through our intuitive knowing and feeling. We can use our creative imagination and emotions to open ourselves to the energy of our heart, while wanting to recognize and realize what it is. As we become aware of the life-enhancing flow of conscious vitality, we can confidently expect to experience its greater reality in our lives.

Enveloping us, there is elevating quantum energy that we can invite into our awareness. If we choose to align our vibrations in resonance with it, we are shifted into its spectrum of energetics. If this is a dramatic shift, such as changing polarities to only positive, it can transform all of our experiences. We can find that we have infinite abilities in how we can use our free will and focus of attention.

Once we have internalized the energy of gratitude, joy and love, we are aligning with the energy of our heart. This is the vibratory spectrum of our intuition and is how we align with it. It is how we can become aware of our eternal, infinite Self. We can free ourselves from our self-imposed limiting beliefs about ourselves and expand into our infinite awareness.

We can be aware of the awareness of all conscious beings, because all live within the consciousness of the Creator and are connected in infinite love. Our intuition has no bounds, unless we impose them, which we have done. Now we can break free and expand to our natural limitless awareness.

Expanding the Presence of Our Awareness

By resolving our limiting beliefs about ourselves, we enable ourselves to be pure present awareness. Without limitations, we have no random thoughts or any anxiety, and we can lovingly allow our ego-consciousness to relax and be silent until called upon. Our free-will-creative ability enables us to live in abundance and freedom. We are free to be loving and understanding in all encounters, even encounters with psychopaths. How we react to others is important to us only because we are constantly creative. The polarity and frequency of energy that we pay attention to is the quality that we are bringing into manifestation in our own experience. It makes no difference if we are welcoming or resisting any energetic patterns. It makes no difference how intelligent or clever or simple we may be in our ego consciousness. We are creating a level of vibration with our attention and how we feel about ourselves. This is what we are expressing in our energy signature. It radiates into the quantum field and attracts resonant energy patterns into our experience.

The more complete we are in being clear, the more open and vibrant our lives become. We can be completely Self-responsible, needing nothing outside of our own creative ability. We become part of cosmic life-enhancement in universal consciousness. We can release our attachments to personal limitations, and we do not need to feel that we own anything. We are free to enjoy and use everything that has come into our lives, but it may leave our presence at any time, to be replaced by something more in alignment with our current state of being. We grow in and out of things and relationships throughout our lives, all dependent on our vibratory level.

There is no one to blame for anything. All of life is a play of energies that arrange their presence in our lives according to our recognition and attention. By paying attention to the energies of our heart and our intuitive receptivity, we can know in every moment everything we want to be aware of. We can be completely without fear, knowing that we are eternally present in infinite awareness. We do not need to think about the past or the future. In every moment we can just be present and aware, always knowing and understanding everything relevant, while we are vibrating at the level of gratitude and compassion.

Our destiny is our own to choose. We constantly choose the polarity and vibratory level of our attention. By being clear and present, we open our awareness to our natural essence. We can become aware of the depth of our Being, our infinite consciousness and creative ability. As we elevate and expand our awareness, we become the masters of the dimension that we inhabit.

Creating a Continuously Heart-Felt Life

In our imagination, we can go to a secret place of beauty and tranquility, where the air is sweet, the birds sing uplifting songs, and we can fantasize that we are light Beings interfacing with other light Beings in wonderful ways. We can make this kind of fantasy into an imaginary experience that we can feel through our senses. In this realm there is only gratitude and joy and deepest love. If we can imagine experiencing all of this, we are taking the first step in creating this way of Being as our reality. It is already in our awareness, because we actually can experience this in our expanded consciousness. The more we intend to realize the light-Being of everyone we encounter, the more we vibrate at the level of the creative life force flowing through our heart. Our experiences are manifestations of our own radiant energies, which we draw out of the quantum field with our state of being, expressed as our personal energy signature. This can be our new world of experience.

One way of creating this kind of vision manifesting into our experience is to release our limiting beliefs about ourselves, so that we can become mentally and emotionally clear, with all personal requirements immediately fulfilled. We add our life force to the light of everyone that we focus on and exchange energies with, including the negative ones, whom we can encounter without fear or engagement, but with compassion and forgiveness, even if they cannot receive it. What matters is our own level of polarity and vibration. By maintaining a vibration at the level of gratitude and love, we modulate the quality of energies that we radiate into the quantum field that envelops us, attracting experiences that reflect the quality of our own energetic signature. This is what we recognize as our reality.

We can realize how we create our own experiences, while at the same time participating in the experiences of humanity. By learning to resolve and transcend beliefs in our personal limitations, we can awaken to amazingly expansive awareness and mastery of our human situation. As we develop deep sensitivity to our intuition by paying attention to our inner knowing and emotional guidance, we find that we begin to live miraculous lives. The energy patterns that could have been difficult or even tragic for us get transformed into positive encounters or distant events. Our experiences result from our own predominant and continuous vibrations and polarity. Once we are aware of this, we can intentionally practice choosing our vibratory patterns by directing our attention in positive ways.

In order to live the most fulfilling lives, we can make achieving mental and emotional control a primary factor in our attention. We can open our awareness to the presence of light beings all around us and learn to converse telepathically with them. When we align ourselves with the energy of our heart, we have so many wonderful capabilities potentially available to us, awaiting our realization.

The Essence of Realization

Realization is a result of our intuitive knowing. It is beyond thought, but it can be stimulated by thought, feelings, encounters, imagination and by our request. When we involve ourselves so thoroughly in a higher math equation or an invention that we haven't been able to figure out mentally, we can retreat into our inner knowing, and the answer will appear in our realization.

We can do the same thing when we are stuck in poverty or starvation. After we have pondered our situation and have not been able to know how to face it, we can go within, open ourselves to the solution we need and wait for the realization that gives us the guidance we need. To realize our inner guidance, we must be in a state of complete acceptance, without any preconceptions and in which our ego-consciousness has given up control, especially if threatened with failure of survival.

Regardless of how difficult life may become, it is helpful to be in gratitude, knowing that we are abundantly cared-for, which we always are, once we can realize this. Realization is the necessary ingredient in making our reality. It is our inner knowing so completely that we are entirely comfortable in its presence. In its truth, there is no possible energetic strong enough to sway it. It is part of the energy of our conscious life force and is able to access universal consciousness. Once we can open ourselves completely to the energies of unconditional love, we can transcend the restrictions we have placed upon our awareness. This may be a moment-by-moment process.

More than awareness, realization makes energetic patterns real in our experience. Realization can be awareness of the energetic patterns that, as humans, we have learned and incorporated into beliefs that limit our realization ability. Our beliefs are like standing waves that have been created by the power of the ocean. They cannot be changed. They can only be resolved and released into the quantum field for the creation of new energetic patterns. Resolution happens in our own awareness through acceptance of the energetic incoherence of difficulties. They're all just experiences that are brought into our awareness for our benefit, guiding us toward our inner Self-Realization.

Becoming aware of our true, expansive Being is what all aspects of life are urging us to do. If we choose to look for these energetic patterns, we can open ourselves to aligning with their vibrations and drawing them into our experience. Once we resolve and release our limiting beliefs about ourselves, we free our power of realization, and we can transform ourselves into Beings of great love, joy and creative power, as fractals of the Creator.

Learning to Be the Masters of Our Lives

As we approach the climax of our sojourn in limited consciousness and creation of negativity, we are ready to awaken to the joy of our eternal awareness in the unconditional love of universal consciousness. We are kept in the prison of limited awareness only by our own choice, but this perspective is deeply set in our subconscious mind and must be released. It can be jolted awake by an out-of-body experience, or it can be trained in deep meditation. It can be transcended with deep breathing techniques, and it can be trained through practice and repetition of envisioning, singing, drumming, bio-feedback and other techniques. With the intention to expand our conscious awareness, we can all find something that works for us.

Only if we engage with the darkness of negativity on its own energetic level are we subject to its effects in our life. Negative experiences can exist for us only if we give them our life force by recognizing them and realizing them as real in our experience. Instead of enduring fear, suffering and lack of fulfillment, we have the option of rising in love and confidence to master every circumstance that we encounter and are aware of.

Once we are able to transcend our immediate circumstances, we can know that nothing exists outside of our own Being, we can realize the power of our own consciousness. The essence of our awareness is unlimited. By controlling our thoughts and emotions in a directed way, we have the ability to modulate the energies in our presence into patterns that are compatible with our visions and feelings. This requires great power of concentration and single-mindedness, which is possible for all of us with practice. It can become our way of being and results in our ability to face every situation with love and confidence, eager to rise to any challenge.

We are essentially Beings of great love and joy, compassion and gratitude, just like our Creator, out of whose consciousness we constantly arise in the miraculous experience of life. We are fractals of the whole of universal consciousness, with the ability to create universes and to enjoy every moment of our eternal existence in sharing and celebrating our exuberant creativity with one another. And so it is.

Journey to the Center of Our Being

Although we are incarnated with a limited sense of who we are, we develop a self-identity as we observe and interact with one another. This is our ego-consciousness for this lifetime. It is the persona that we have created and can change at any time by shifting our realization about ourselves. This most easily happens when we are able to be in the border of waking and sleeping. Here the ego is silent. Nothing is happening for it in our attention. We can be pure presence of expanding awareness. Individually we can feel connected with the consciousness of humanity extending into infinitude.

Our natural awareness is everywhere. We have no limits outside of ourselves, and we can have control of our own limits, allowing those necessary for being an incarnated human, and concurrently opening to infinite consciousness. The vibration at this level of consciousness is unconditional love, joy and absolute confidence in the reality of our heart's guidance.

Being in this calm and easy space of deepest love and creative manifestation, we can embrace ourselves and all beings with unconditional love and compassion for all misdirected energies. As we vibrate at attractive positive energy waves, maintaining love and joy, our perspective can be based in deep understanding and emotional sensitivity. We can change the perspective of any negativity that may be present by holding only love for the luminous essence of all beings we encounter. We can be aware of all of the energies that are present and their qualities. If we want to engage with any of them, we focus our awareness on their vibrations and align with them, or we attract them to align with us.

The requirement for being able to live at a high level of consciousness is being able to shift our realization to what we truly want. It begins with our intention to be alert at the vibratory level of ego stillness. Nothing of ego-consciousness that is negative is meaningful. It is our own creation, and we can resolve it with acceptance, compassionate understanding, and then release it. To be the masters our lives, we can shift our focus to the vibratory level that we want to experience in the greatest love that we can imagine.

With sufficient practice, we can realize its reality, and we begin to transform into our true Self of infinite Being. This is our destiny, and we are expanding the consciousness of humanity.

Being the Transformers of Humanity

Divine consciousness fills us and is all around and within us, and we have the free choice of being aware of it, or blocking our awareness of it by believing in our limitations. Occasionally we may get glimpses of it, when we find ourselves in a dire situation, and we open ourselves to divine intervention. Then something happens, and we end up in an acceptable state of being, thankful for our survival. If we desire to be aware of the greater consciousness that we all share and that draws on the experiences of all of us, we can accomplish this by intending to be aware of it in ourselves.

When we become acutely aware of our inner knowing, we find that we are being drawn beyond our limiting beliefs. Life becomes a moment-by-moment adventure, as we begin to follow the prompts that our intuition uses to guide us into greater love and joy. When we choose to be aware of the divine consciousness within us, we automatically know how to be in every situation. While existing in the eternal presence of Self-Awareness, we can recognize our expansive Being beyond time and space, expressing ourselves as humans in the time/space empirical world of duality.

Once we transcend our limiting beliefs about ourselves, our lives are transformed, and our experiences become life-enhancing in every way, because this is the energetic expression of universal consciousness, always expanding into infinity. We can become masters of human life, beyond karma and suffering. We may live an ordinary life, but from within we radiate gratitude and joy. We become radiant with inner light and are helpful with others in every encounter.

We are the creators of a better world for ourselves and all of humanity. The chaos and suffering in the current world will stop when people stop aligning with it. It needs our life force to exist. As the consciousness of humanity expands, there is a growing recognition that we can intentionally align with life-enhancing ways of being. This brings peace among us, and we can learn to work together as one species, arising out of the consciousness of the Creator of infinite Being.

As our radiance expands, others notice. Our lives become experiences of deeper love and abundance. We become transparent in everything, while living in the light of our heart. Guided intuitively in every moment, we naturally create everything we need and want in alignment with universal consciousness. We are the transformers in the energetic circuits of humanity.

Realizing the Truth of Our Being

We are enveloped in the energetics of abundance, freedom and joy, but most of us have chosen not to realize this. It hasn't been a conscious choice, because we have been extensively programmed to keep this realization out of reach; nevertheless, the dimension of fulfillment is always available to us. It is in our lives now whenever we can realize it. It is the destiny of humanity to ascend into realizing unconditional love and beauty within and all around. Those of us who know this have the task of maintaining this level of consciousness while interacting with others.

By our radiance we are shifting the consciousness of humanity into positive, high vibrations of gratitude, compassion, appreciation and deep understanding. We have learned that our primary task is shifting our own awareness into positivity in every moment and being open to our intuitive knowing. Once we achieve a higher state of consciousness, we radiate this into the consciousness of humanity, and it reaches everyone to the extent that they are open to it.

As more of us realize our eternal, infinite Self, it becomes easier for everyone else to become aware of their true Being. We are being carried in this direction by our natural environment and the elevating energetics that are being directed toward us. It is becoming easier to live in gratitude and joy, while it is becoming more difficult to live in fear and greed. The light is growing stronger, and we can realize it within. When we do this, it elevates the energies around us and in everyone we interact with.

It is becoming apparent that the most powerful force in existence is the unconditional love of our Creator. It is so powerful that it destabilizes negative energy and negatively-oriented beings, and ultimately demolecularizes all manifestations of negativity. Their energy returns to the Source of all for transformation. This is the energy that is drawing us into alignment with the intuitive energy flowing to us through the heart of our Being. It enables us to transmit the radiance of our alignment with truth and unlimited Being.

Whether we realize it or not, we are the creators of the quality of our personal lives, and our transformation must come from within ourselves. Being able to realize our true essence as infinitely powerful creators in unconditional love, allows us to transform the consciousness of all who are open to it. We can begin to know one another as great Beings of Light, while we go about our daily lives, following our intuitive knowing and always being attracted to more light and more perfect alignment with the guidance flowing through our intuition.

How We Create Our Reality

We can ask within for guidance in opening our awareness to the greatest love and joy, which comes to us when we can realize its reality. To realize this reality, we must ask for it within ourselves. When we want it and can create the experience in our imagination and emotions, we are guided to our fulfillment. It happens when we realize it. Without our realization, it cannot exist for us. Quantum physics has proved that, by interacting with the energetic patterns that we recognize, we create their reality for ourselves. When we do not recognize them as real, they do not exist for us, because it's all just swirling energy, and we are the modulators of the energetic patterns that we interact with.

Another realization, that is related to knowing our own reality, is that our empirical experience in duality is an artificial creation held in existence by the imaginary and emotional life force of humanity. We create reality for ourselves consciously and sub-consciously. Without our recognition, it would not exist for us, but it would for all others who recognize it as real. In order to participate in it, we must exist within its limitations, but we do not have to limit our awareness to this compartment of consciousness. By aligning with the positive, high-vibratory energies in the heart of our Being, we can live fulfilling lives in every way.

The divine flow of our conscious life force is limited only by our deeply-buried limiting beliefs about ourselves. These were developed by our ego-consciousness, which also trained our subconscious, where our inherited beliefs reside. The existence of our ego-consciousness depends upon maintaining our limiting beliefs. Without these limits, ego consciousness melts into our full awareness, guided by the inner knowing that constantly comes to us through the conscious life force in the heart of our Being.

Opening access to our inner knowing is crucial for further realizations. We know what our conscience is, and it is part of our inner guidance. If we can align ourselves with the vibrations of our conscience, that is the beginning of inner knowing, and it will lead us deeper into greater awareness of higher vibrations. As we open our awareness, we can recognize the reality of the realm that we vibrate in resonance with. It becomes our reality. Every realm exists here for those who realize it's reality, and it exists contemporaneously in the same space as human empirical duality. It's just a different realization of reality. It may appear to be the same as the empirical vibrations that we know, but in the realm of only positive polarity, there is no duality.

It does require a leap in consciousness to open to the reality of a different level of awareness. That leap is resolution of our limiting beliefs, by developing sensitivity and alignment with our intuitive knowing. We can begin to transcend our beliefs by finding them unbelievable. The leap requires learning to use our recognition and realization in creative ways that may seem unrealistic or impossible. If we can intentionally transcend these beliefs, we can understand their basis in fear and doubt. These do not exist in the higher vibrations beyond duality. There is only the vibratory level of gratitude, joy, compassion and their companions.

Our Place Among Celestial Beings

What do the stars hold for us? Each of them has a person-hood, just like our Sun and Gaia. The moon is artificial and has no consciousness beyond its structure, but because of its gravitational and magnetic pull, it influences everything liquid in us, including our emotions. We can get to know the stars and planets as expressive, living beings, just like us, but without ego limitations. The various astrological systems have given us rudimentary descriptions of the character of our Sun and each of the planets in our Solar system, but the character of the stars is largely unknown, although some indigenous people know the nature of their home star systems. All of these celestial beings, with their unique energies, influence everyone, and we interact with them in various ways.

Stars are the most powerfully radiant beings in our dimension. They provide warmth and can fill us with joy. Our planets are less evolved, but because of their personal magnitude, they are still powerful influences on us. Due to the way divine geometry works, there are critical angles of their location relative to us that cause more benign or challenging aspects for us in the ways that their energies come to us. All of these celestial beings are stimulating us to evolve in greater ways in expanding our consciousness.

We live in an expanding universe, and the natural energy expresses greater creation and enhancement of all life. As our Sun has evolved from yellow to white, he is brightening our lives with greater warmth and expansiveness. The entire cosmos is experiencing rising vibrations and is carrying us along with it. Our mother Gaia is growing more resplendent and is regenerating in many ways, as we have begun to become more aware of her needs and abilities.

As we immerse ourselves in the energies of nature and all of her life forms, we can expand our awareness in alignment with the rising vibrations of the Earth. We need to pay attention to all of the challenges that Gaia is giving us, because she is clearing herself of negative energy. There has been much evil manipulation of the weather, unusual floods, gigantic fires and even earthquakes, but the manipulators no longer have access to these technologies. It is all ending, and they are leaving the planet.

It is time for us to celebrate the rise of love and joy in all of life and align ourselves with these energies in all aspects of our lives and in all our encounters and relationships, beginning with our own being. We have been disguising ourselves as humans, when in fact we are great celestial beings ourselves with unlimited potential creativity and infinite awareness. Only our limiting beliefs keep us from realizing our true nature. We are now free to open ourselves to our potential and shine like the stars.

We Are the Force of Light on This Planet

Eons ago we, the human species, planned, designed and manifested the empirical world of duality in order to expand universal consciousness into a realm that had been impossible to experience as beings of eternal, infinite awareness. We designed a realm where we could realistically experience a feeling of separation from our true Being. Since that time, we've been through the entire spectrum of duality, collecting a myriad of negative experiences, as well as wonderfully sensuous experiences. We got so entranced in this realm, that we've allowed ourselves to become enslaved to negativity in fear, doubt and much suffering. Now our species consciousness has decided that it's time to awaken to our true Being and move beyond the limitations of empirical duality, in alignment with our entire galactic congregation.

There is a leading portion of humanity who have become aware of what is happening on a galactic level and are raising the vibrations of humanity, so that all may become aware of our greater reality. Those of us who have remembered experiences beyond the physical body have awareness of our limitless consciousness and boundless possibilities of expression in unconditional love and joy. By intentionally living in the level of vibration of these realizations, we are radiating silent and powerful energetics throughout all of humanity, and we are having a dramatic effect.

Because we are all individualized conscious beings, each of us must come into our own realization of the truth of our Being. Even though we are part of a group consciousness of an entire species, we are also individuals, and we have our own realization of our own reality. Each of us has our innate way of realizing our reality, and it comes through our inner knowing. When we begin to search within for this awareness, we can begin to notice that everything in our lives is symbolic of our state of consciousness. It is all waiting for us to be in love, gratitude and joy for all of it. When we do this, the negative polarity that we've been oriented to shifts to positive, and everything changes in our experience.

Our breath becomes deeper and more resonant with greater consciousness. Our presence becomes confident and powerful. Our stature becomes more erect and beautiful. Our poverty becomes abundance. Our physical defects become vibrant health. Our ignorance becomes brilliant awareness. Our enslavement becomes freedom and sovereignty. This all takes practice and dedication to making the necessary changes in our personal awareness, and this is the miraculous path to higher consciousness and awareness of our true Being.

Empowering Ourselves to Achieve our Potential

We live in an electromagnetic dimension of positive and negative energies. The positive energies are life-enhancing, and the negative energies are life-diminishing. They are mirror images of each other flowing together forever creating a neutral essence. In our essence we contain the potential to manifest both negative and positive forms. It is our choice which of them we choose to pay attention to. By our attention and alignment, we create experiences for ourselves of the quality that we focus on and realize as reality.

By focusing on events and beings that we are angry or fearful about and want to defeat, we disempower ourselves and all of humanity. We are focused on the negative, when the positive is just as much potentially present. It just requires our attention and realization. How do we know when we are negatively or positively oriented? We know it by how we feel. We know when we are anxious, stressed or angry, just as much as we know when we're thankful, compassionate and joyful. Every present situation has the potential energy of either charge. It is our choice of which we decide to pay attention to and align ourselves with.

The world that appears to be outside of ourselves is actually a reflection of our own consciousness. What bothers us out in the world is a manifestation of our interior perspective and the quality of our own realization. This is what we can transform by choosing to realize the positive side of every event and happenstance, and every person we encounter. The light is ever-present. It only needs our realization.

Once we understand the nature of our own reality, we can master our empirical world of duality. The negative cannot exist without the positive. The secret is that the positive can lead us beyond polarity into a higher dimension, where the negative does not exist. This is our destiny as great Beings of Light and unconditional love in the fullness of divine Being. This is our true nature beyond the world of duality.

We Are Being invited to Expand Our Awareness

Each of us is an expression of a higher Being, a Being beyond polarity and beyond comprehension. Our connection is through the energy of the heart of our Being, expressing itself through our intuition, our inner knowing. If we pay attention to our inner knowing, it becomes clearer. In every moment our knowing is present. Through it we can understand everything that happens in our presence and awareness. Our presence is our electromagnetic expression of our conscious state of being. Our awareness lives throughout universal consciousness, wherever we guide it with our attention.

We're being invited to transcend our limiting beliefs about ourselves. An example might be the belief that we need to make money to survive in our society. This is an ego-expression of limitation. We already have everything we need and could ever want in expressing the energy of the heart of our Being. We have complete fulfillment available to us however and whenever we choose, or even constantly, in the present moment.

In the dimension of time and space, there is always a moment, one after the other. It is a moment apart from time, and it is when our intuitive knowing happens. It is our connection with the eternal aspect of our Being, but our limiting beliefs have kept us largely unaware of it. In order to gain infinite knowing, we need to be in the moment beyond time, the ever-present presence of clear awareness. That means practicing being clear, without attachment to limitations, while resolving all fear.

Although all that is required of us to realize our infinite and timeless Self is complete openness and receptivity to expansion, we seem to find it challenging to achieve this state of being, because of our limiting beliefs. We can realize what our limitations are by feeling the quality of their energy. We know what any negativity feels like. There's at least a tinge of fear, ultimately based on belief in our mortality. All limiting beliefs are based on negativity. They diminish our life. When we're experiencing our divine Self, everything we need and want just comes to us. We attract everything that is positive and life-enhancing for all.

We are already fulfilled in every way, just by our state of Being. Universal consciousness manifests what we create, now and always. It is our own state of being that forms our energetic transmissions out through our aura, attracting and being attracted to resonating energetic patterns that manifest as our empirical experiences. Living in this dimension becomes much more fun, when every moment is wonderful. This happens for us whenever we realize Who we are, and we can embody our divine energies.

Living Joyfully in the Present Moment

We are all the same Being, the same Consciousness of the Creator, whose creative life streams through us constantly in as much abundance as we allow to flow through us. We are our own filters of universal consciousness. In our essence we are limitless awareness with awesome capabilities. All of universal consciousness flows into us now and always in our own awareness, which manifests as our intentional personhood in this life. We have been gathering all of the experiences that we have wanted and needed to complete our earthly sojourn in duality, and when we decide to open ourselves to our eternal, infinite awareness, we have that choice. By commanding it to come into our experience, and maintaining the perspective of mastership, we can attain awareness of our true Being. It requires a strong intention to transcend all limiting fear-based beliefs about ourselves and to declare their resolution in deepest love and gratitude.

It is our option to change our polarity and raise our vibrations. Our personal truth in Being is always present in us, and when we can achieve mental and emotional clarity beyond limitations, it can be in our awareness. When we can recognize the reality of our eternal, infinite awareness, we are no longer subject to limitations of any kind. They become unbelievable in reality, and nothing keeps us from living in the present moment in gratitude and ecstasy, guided by the intuitive knowing of our heart. We cross the threshold of dimensions, and our world changes to one of miracles and amazement.

When we expect to have to interact with negativity in the

world, we create this kind of interaction. The way out of negativity is to change our perspective and orientation to one of constant gratitude and joy. Nothing beyond ourselves keeps us from being in a state of absolute confidence in, and identification with unconditional love and unlimited awareness. This is our personal choice at any time. When we make that choice with absolute intention and confidence, we open ourselves to living in an energetic dimension beyond duality, and our experiences become wonderful and fulfilling.

By transcending our ego-conscious limitations through fearlessly following our inner guidance of heart-consciousness, we enable ourselves to master every situation in confidence and compassion. Our personal needs become our fulfilled experiences in a higher-energetic dimension. Once we have transcended fear of mortality, our participation in any energetic level of vibrations is voluntary on our part. When we can realize our immediate creative ability, we can be free of all bindings, attachments and limitations. We can realize our reality in a truly wonderful dimension of living, and we can experience loving and truly entertaining relationships, guided by the intuition of our heart.

Continuing a Deeper Understanding of Life

In every moment we can receive complete realization of the infinite essence of who we are. Our potential abilities and conscious realization are so far beyond the apparent limitations of humanity, that we have been unable even to imagine eternal, unlimited awareness and infinite creative ability for the enhancement of all life everywhere. Flowing to us always is the conscious creative life force that is the essence of the universal consciousness that all living beings participate in, and that endows us with constant creative power, expressing itself in energetic alignment with our state of being.

By our predominant thoughts and emotions in every moment, we express the polarity and vibratory frequency of our creations. By realizing ourselves as the humans that we have been taught to be, we have believed that we are mortal and subject to a higher power. If we thoroughly examine this realization, it becomes unbelievable. Energetically, we are expressors of consciousness. Our expression radiates from the essence of our awareness in the form of electromagnetic waves having our individual polarity and frequency in the way that we modulate them through our imagination and emotions, within the energetic boundaries that we impose upon ourselves.

Through our awareness and interactions with the world that we make real for ourselves through our realization, we participate in a play designed to guide us to awareness of our inner knowing. When we engage with the energetics of duality, we remain confined within the spectrum of vibrations of this realm. Fighting against energies that we dislike or feel threatened by holds us within the polarity and frequency patterns of those energies and continues our creation of this spectrum of energetics in our experience. We can ask ourselves what we are afraid of. This is the basis of our limiting experiences and the essence of our ego-consciousness.

We can allow all energies of every kind to exist, those we love and those we feel challenged by. We do not need to seek anything out or try to destroy or avoid any of them. We can just be present and aware of all potentialities. This ability is within our own consciousness, and we can open ourselves to its realization. We don't have to seek it. We can just be present in clarity and allow the fulfillment of our Being and awareness of greater energetic dimensions to come into our realization.

We can sense what is enhancing all of life. In the heart of our Being, we know what this is. It is called unconditional love and creative life force. By aligning with this energy, we become our true, expansive Self, unlimited in every way and possessing mastery of every situation that comes into our experience.

Chapter 6. Living in Unlimited Awareness

Expanding Our Transformative Powers

As humans we have the opportunity to realize our eternal, infinite awareness and unlimited creative ability. Our human persons are only a masquerade for purposes of experiencing duality as if it is real and to enjoy the physical experience. Beyond the limiting beliefs that we have imposed upon ourselves to create the reality of human life, we can awaken into a completely fulfilling awareness, encompassing the awareness of every conscious being everywhere in unconditional love, compassion and joy. We can know and experience that in our essence we are free spirits, unlimited in whatever we wish to make real with our realization.

In our full awareness, we are multidimensional and can express ourselves as any being we wish to experience, and we can identify with the awareness of anyone. We are all the same essential conscious life, sharing our eternal awareness with everyone. The world of relativity, as Einstein understood it, is a synthetic creation of human consciousness that is dissolving into awareness of quantum expansion, in which the reality is a result of our personal recognition and realization, interacting with energetic expressions.

Humans appear to be fragile and necessarily fearful, lest we be swept up into chaos and suffering or be terminated. This is all a trick we play on ourselves in our consciousness with our limiting beliefs about ourselves. These are all artificial constructs that we have imposed on ourselves and can resolve at will. Because of the fear-based, limiting beliefs on which our ego-consciousness is based, we have been afraid to open ourselves to the truth of our Being. We have been unwilling to believe who we truly are and have not even wondered about our expanded identity.

In the vastness of our consciousness, we are the creative masters of universes and beyond, if this is the desire of our heart. When we are open to our intuition, we can realize the guidance coming through our heart, giving us infinite creative ability in every moment. In our true Being, we are beyond fear and doubt, beyond time and space, eternal in our infinite awareness within the consciousness of the Creator, of whom we are fractals of infinite creative Awareness, constantly and forever flowing divine life force into creative expressions through our moment-to-moment recognition and realization.

Designing and Implementing Our Destiny

When we can expand into full consciousness of our true identity as our infinite presence of awareness, we still have the choice of interacting with those who are still entranced in the realm of dualistic empiricism and limited awareness. This kind of interaction can free humanity from enslavement to negativity. When we are guided by the energy of our heart, we can voluntarily participate in the realm of duality from an awareness of unconditional love and compassion.

Humans are designed to be free spirits. We can only be enslaved if we allow negativity to control our attention through the formation and belief in personal limitations. We are limited only by our own chosen characteristics of awareness. We have voluntarily created our needs and defects by life-diminishing thoughts and feelings. These are choices that we can learn to transform in creative ways.

Every experience that we create results from our nature as fractals of Creator consciousness. By our vibrations, we are constantly creating energy patterns that manifest as the qualities of our experiences. By aligning our emotions and imagination with the energetics that we choose, we have the power to create the vibratory level of everything about ourselves and our interactions.

By feeling ourselves into a state of gratitude, compassion and joy for no reason other than to experience their vibrations, we can accustom ourselves to living in their energetic spectrum. These high vibrations can carry us miraculously through any confrontations with negativity. As we confidently radiate high vibrations by feeling that we are in a high-vibratory state of being, we draw the consciousness of those around us into our energetic spectrum. Our heart-centered state of being destabilizes negativity in our presence and raises the vibratory level of our environment in alignment with the Spirit of the Earth.

By living in gratitude, compassion and joy, we open ourselves to the power of our creative essence. We have the ability to transform any situation through our powerful heart-centered vibrations. In full access to universal consciousness through our inner knowing, we can live with life-enhancing creativity in every moment in limitless ways.

To begin to describe the kind of transformers we are, we could compare it to transforming a 12-volt system to a 50,000-volt system. And even more, because we are unlimited in what we can do with our consciousness. It is all purely our choice of how we want to live. We are completely in charge of our destiny.

The Light of Our Life

We are beings of living light, emitting photons, which are quanta of light. They are conscious, living beings expressing themselves as electromagnetic wave patterns that appear to us as light. They exist within us and all around us. They are fully aware of our essence and interact with us in ways that enable us to perceive them. When we choose to focus on them, we see them. When we do not realize their presence, they still exist in their own being and express themselves energetically, but they do not appear in our awareness or in our reality. They have no mass and no limited form.

Photons are multidimensional beings, just like us. They have an essence that is beyond time/space, and, like us, they can express themselves within the empirical world. They are given to us with our conscious life force that comes to us constantly from Creator consciousness. Their essence is the same as ours. As fractals of the divine One, we are connected within the same consciousness. We can always be aware of our inner light, and we receive as much as we allow. The more photons we emit, the brighter we appear.

As we resolve our limiting beliefs about ourselves, we can become more aware of our own access to infinite consciousness. Along with a greater sensitivity to our inner knowing, we can become aware of greater light in our own being. We can choose to have a desire and intention to enhance the life of everything and everyone in our presence. In this process, we naturally enhance our own lives without even having to be mindful of it. We can create the highest energies of love and joy to manifest for us, because we can align ourselves with the energy of our heart. This draws other heart-felt beings into our presence, along with abundant living and personal freedom to express our intuitive guidance.

Once we are fully aware of our intuitive guidance, and we can express ourselves in alignment with it, we become creators of wonderful experiences that we are deeply grateful for. We can receive the unfettered flow of conscious life force, radiant with Source Light flowing through our heart and into our awareness in joy and love. By keeping our attention on positive scenarios, we continue to create fulfillment for ourselves and all around us.

Creating More Joy in Our Lives

Because of decisions that we have made consciously and subconsciously in our essential personal being, or soul-consciousness, we have been born into our conditions and circumstances. We are the only ones who can create the quality of our experiences. If we believe that we are poor and barely surviving, this is what we realize as our reality. Entire communities have this kind of belief system. If we change our focus to believing we are living in the abundance of everything, that is the condition that we create in our lives. The challenge is to be able to change our beliefs from believing in limitation to believing in abundance for all. This requires a close examination of personal beliefs about ourselves.

In order to live in great joy, we can train ourselves to realize the light in everyone, beginning with all of humanity. We all have the desire for joyful, abundant living, and we can all experience it by knowing that we are living in its energy. If we were created to live in lack, we could never achieve anything better. But we do not want the condition of lack. It's not normal for us. What is normal for us is what everyone feels in our deepest consciousness. That is where abundance lives. We can make our way to this realization.

There are many ways of doing this. We can imagine what it would be like to live in every kind of abundance. Beyond ego consciousness, there is no money, because money is currency. It is the manifestation of the flow of energy. In our innate self, abundance manifests in everything and in every experience. Money flows with everything else and appears from many sources, previously unknown to us. We naturally have more than we need in every way. To ego-consciousness, this is impossible, and it can only happen by constant miracles.

Welcome to quantum physics for some basic understanding of the workings of our reality. Everything that exists for us, is a plasma environment of conscious entities expressing themselves as electromagnetic waves and wave patterns, charged positively or negatively and having amplitude, frequency and wave-length. These are all expressed within universal consciousness.

There is only one consciousness, which we participate in according to our state of being, which is a result of the polarity and frequency expressed by our predominant focus of awareness, defined by our beliefs about ourselves. When we focus on the vibratory pattern of any entity, we interact energetically with that entity, and we pick up the resulting vibratory patterns that stimulate our senses and our subconscious awareness.

Once we realize that our deepest desires arise from the essence of our consciousness, which is the consciousness that we all share, we can also realize that we have a living conscious connection with one another. We have natural telepathy and much more. Our access to these abilities can come from resolving our limiting beliefs and opening ourselves to the realization of universal consciousness. This is the realm of joy and everything that feels good and wonderful for everyone. By opening our awareness to this level of consciousness, we can create its reality for ourselves. In the radiance of this consciousness, we brighten the lives of everyone.

Intentionally Transforming Our Destiny

We have allowed ourselves to be trained and conditioned to enslave ourselves to negative imaginings, resulting in our suffering, physical deterioration and abuse of ourselves and our planet. It is all an imaginary creation that we have kept recreating. Our only enemies are within our own consciousness. They exist only because we provide them with our conscious life force through our mental and emotional abilities. Everything we experience is a result of our own creative energies. Because of who we are, we can change everything through our intentional choices in focusing our attention and awareness.

When we learn to control the focus of our attention, and we are aligned with our intuitive knowing, we effortlessly create miracles and transformative experiences. We can do this by going with the flow of the energy that runs through the heart of our Being. We can know this flow of conscious life force coming to us in every moment by how it feels emotionally. Because we have experienced their qualities for eons, we naturally know the difference between negativity and positivity. We have the power to choose which one we want in our awareness.

Because it is based in fear and negativity, our ego-consciousness consists of many defects, which manifest in our physical, mental and emotional selves. Since they are all self-created, we do not need to be entranced and enslaved by our defects. They depend upon our belief in them as real. Once they become unbelievable for us, we are free to transcend their energetic patterns and clearly know our true Being as our eternal presence of awareness with infinite creative ability. This process happens through our realization, as our awareness expands, unimpeded by limiting beliefs.

There are many ways to make this shift in our conscious awareness. We can work with our consciousness by intentional focusing on what we truly want with gratitude, joy and feelings of fulfillment. Because they are beyond limitation, being in the spectrum of these energies transcends negative beliefs about ourselves and our condition. By transcending doubt and fear, we can give our attention to what we deeply feel and know within ourselves as our unconditionally-loving and life-enhancing gratitude and joy. This is the vibratory level of the intuition coming through our heart and enabling us to connect with universal consciousness.

When we choose to stop giving our life force to life-diminishing feelings and thoughts, and instead focus our attention on our deepest knowing in life-enhancing ways, we can train ourselves to live in the realm of gratitude, love, compassion and joy. To achieve this, we can pay attention constantly to the light and vitality in every being. When this level of energy fills our awareness, negativity disappears from our lives. Our bodies regenerate with the fulness of our life force, because this is how we realize our essence to be, and our relationships become clear and free. As many of us do this, it transforms the consciousness of humanity.

www.ingramcontent.com/pod-product-compliance
Lightning Source LLC
Chambersburg PA
CBHW070632160426
43194CB00009B/1438